New Guide to Restorative Drying

New Guide to Restorative Drying

Revised Edition

Dri-Eaz Education Series – Burlington, Washington

Dri-Eaz Education Series, Burlington, WA 98233
 First edition 2006
Revised edition 2006

Printed in the United States of America

ISBN-13: 978-0-9776701-1-6
ISBN-10: 0-9776701-1-2

Dri-Eaz Products, Inc.
15180 Josh Wilson Road
Burlington, WA 98233
(360) 757-7776
www.dri-eaz.com

Table of Contents

Drying Science — Chapter 7

Drying Equipment — Chapter 8

Drying Systems — Chapter 9

Microbiology — Chapter 10

Preface

DEDICATION

This book is dedicated to the further advancement of the science and practice of water damage restoration to benefit both the industry and its clients.

DISCLAIMER

Dri-Eaz Products, Inc. and any other contributors to this publication expressly disclaim, and shall not be liable for, any and all damages of any nature whatsoever, whether direct or indirect, arising from or relating to the publication, use of or reliance on the information contained in this book or on the CD-ROM, including without limitation any and all special, indirect, incidental, compensatory, consequential, punitive or other damages (including damages for personal injury and/or bodily injury, property damage, loss of business, loss of profits, or the like), whether based upon breach of contract, breach of warranty, tort (including but not limited to negligence and gross negligence), product liability or otherwise, even if advised of the possibility of such damages. The foregoing disclaimer of liability and damages is a fundamental element of the reader's use of the information contained in this *New Guide to Restorative Drying* and this document would not be published without such limitations.

NOTE TO INTERNATIONAL READERS

The *New Guide to Restorative Drying* is published in the U.S., so references, terms, and measurements primarily reflect U.S. practices and standards. Wherever possible, metric measurements are listed, usually in parentheses after the conventional U.S. measurement. Alternate terms are included in parentheses after the first usage of the U.S. term. Depending on the country, other terms or standards may apply.

All health and safety legislation referenced or quoted in the Guide is based on U.S. law. It is the reader's responsibility to become familiar with and operate according to the legal standards in other countries as applicable.

REFERENCE SYMBOLS:

- Notification of web reference
- Important lesson
- Located in glossary
- See appendix and/or CD-ROM
- CD-ROM

Dri-Eaz Class Details

Dri-Eaz offers classes based on the content in this book. For details, go to www.dri-eaz.com/Classes.

DRIEAZ EDUCATION
·FIRST IN DRYING SOLUTIONS·

Contributors/Acknowledgments

AUTHORS

Brandon Burton, Dri-Eaz Products
Kevin Fisher, Dri-Eaz Products

DESIGNER

Michelle Ramlow, Dri-Eaz Products

EDITORS

Marcia Neu, Dri-Eaz Products
Jennifer Ranz, Dri-Eaz Products
Courtney Shannon, Dri-Eaz Products

Special thanks to the following for editorial and graphics assistance: Josh Gates, Michelle Malmstrom, Catherin Nordmark, John Ormsby and Kevin Wolfe, all of Dri-Eaz Products.

REVIEWERS

Dri-Eaz thanks the following individuals for reviewing the book and providing critical feedback to ensure the accuracy, clarity and usefulness of the book to water damage restoration professionals, as well as to other stakeholders from the insurance and construction industries.

Peer Reviewers

Keith Desserich, RestorAid, Cincinnati, OH
Pete Duncanson, ServiceMaster Clean, Memphis, TN
Darren Foote, Water Extraction Technologies, Sparks, NV
Ron Gray, Barker Hammer, Minneapolis, MN
Lew Harriman, MoistureDM.com, Portsmouth, NH
Bob Lintzenich, Servpro, Gallatin, TN
Rob Smith, Professional Carpet Cleaners Supply, Euless, TX
Ernie Storrer, Injectidry, Kirkland, WA
John Straube, PhD, University of Waterloo, Ontario
Paul Thompson, Town and Country Service, Greenland, NH
Cliff Zlotnik, Microban Systems, Inc., Coraopolis, PA

Dri-Eaz Instructor Reviewers

Cara Agerbeck, Gee Plus, Elgin, SC
Richard Driscoll, Gee Plus, Winnsboro, SC
Darren Hudema, Dri-Eaz Products
Nigel Hewitt, Dri-Eaz Products Ltd.
Ken Miller, Dri-Eaz Products
David Oakes, Captain Clean, Sheridan, WY
Richard Ruch, RKR Instruction and Inspection Services, Broomfield, CO
Steve Swan, Dri-Eaz Products

Water is considered the earth's most precious resource. It accounts for approximately 70% of the earth's surface, has strong bonding properties and the ability to dissolve materials. Water is essential for life.

At the same time, water is the single leading carrier of pathogens in the world today – this potential for serious adverse health affects should never be underestimated. In Mexico, one person out of every 60 dies of waterborne disease (cholera). In Africa, a person dies each minute from such disease.

Water is also the cause of damage in some of the worst natural catastrophes North America encounters on a regular basis: hurricanes, which account for eight of the 10 most costly catastrophes ever recorded in North America. In 2004 hurricanes caused $21.47 billion in damage. In 1992, Hurricane Andrew alone caused over $26.5 billion in damage. Andrew's record was broken in 2005 by Hurricane Katrina, which caused over $125 billion in damage, as reported by *USA Today*.

As phenomenal as hurricanes are, they are not the leading cause of water intrusion and water damage in structures today. Often excessive moisture exists in buildings because of construction defects, plumbing failures, damage from freezing and even improper maintenance. According to *Claims* magazine, freezing pipes caused 16.85 percent of average U.S. homeowners' property damage claims in 1997.

Clearly, excessive water has the potential to cause many problems in the built environment. The elements necessary to support microbial growth are always present; all that is required to initiate a destructive chain reaction is the addition of excess moisture. Microorganisms will rapidly flourish and cause structural damage. These microorganisms, in turn, raise serious health issues for building occupants. When a water loss occurs, the single most critical factor in defining the amount of potential damage lies with the amount of time the structure remains abnormally wet. It is for this reason that, initially, the most critical course of action in restorative drying is to respond quickly. All other lessons covered within this manual pivot upon this very important principle.

Restorative drying is a process used to recover from many forms of water intrusion by assessment, documentation, controlled drying and selective replacement. Although there are scientific principles applied throughout the process, much of what the restorer encounters is influenced by a large number of variables. Due to the number of variables, the restorer must frequently revisit the tools, systems and science throughout the drying process.

The goal of the restoration process is to transform an abnormally wet, potentially damaged structure into an environment that can be reoccupied, of equal or better appearance and cleanliness than before the intrusion occurred, and to do so in the most economical and efficient means possible.

The restorer begins this process by identifying all affected materials.

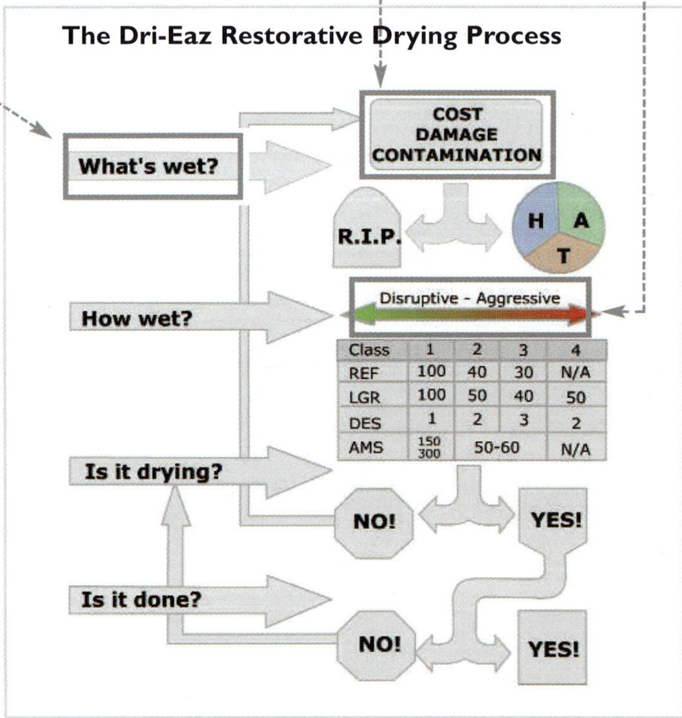

What's wet?

Water must be tracked from its initial source and diligently followed in every direction it migrated to establish an accurate perimeter. Once the perimeter is identified, the restorer then documents what types of materials have been affected (e.g., type of subflooring, type of carpet and pad [underlay], type of wall construction, type of insulation).

Affected materials are evaluated against three criteria to determine if they should be restored or replaced.

The three criteria are:

COST
DAMAGE
CONTAMINATION

Only after identifying the material and considering these three factors does the restorer decide which materials should be dried and which should be replaced. In addition, the restorer will also decide which materials should be dried aggressively and which should be dried through more disruptive means.

Disruptive — Aggressive

Materials of higher value that do not show signs of permanent damage and are either uncontaminated or are easily decontaminated are dried aggressively. This process is also referred to as drying in place, meaning that the material is dried without being moved or manipulated in any way.

Materials that have suffered damage, are contaminated and cannot be decontaminated, or are of relatively low value are removed or manipulated to facilitate restoration, rather than attempting to dry these materials in place. This process is referred to as disruptive, and may include the use of inter-air drying systems, perforation, removal of surface or finish materials (e.g., vinyl sheet goods, wallpaper, wallboard, etc.).

All materials that are determined restorable are inspected further, and the amount of moisture absorbed by the material is measured and documented. Answering "How wet is it?" **How wet?** is critical to the drying process for two primary reasons. This information can influence the amount and type of equipment used. It is also necessary in order to determine if drying progress is being made on subsequent monitoring visits.

Drying progress is noted during daily monitoring visits. The restorer asks "Is it drying?" while comparing two sets of data. The first and second sets of data, recorded on a monitoring report, reflect the impact of the drying effort on the **Is it drying?** affected materials over a 24-hour period.

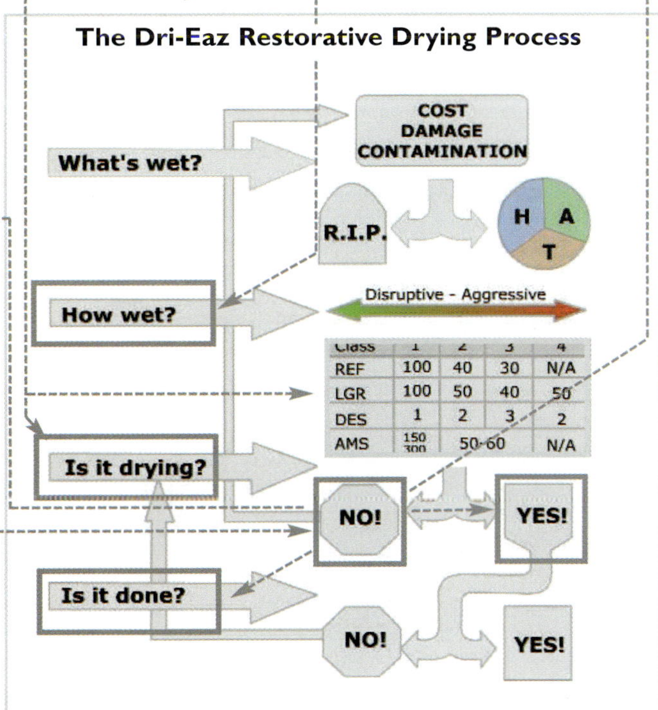

If materials are making progress **YES!** after the initial 24 hours of drying, then drying continues; no major change is necessary to the process. The restorer continues to observe progress until **NO!** drying is complete.

If materials are not drying the restorer must re-evaluate the equipment, systems and processes. The current system is not working. Change must occur. The change can be in the form of being more aggressive, or more disruptive. The evaluation process is identical to the initial assessment except in one important detail. The material has been wet for an additional 24 hours. Damage and contamination are of greater concern.

Many restoration projects will remain in this cycle of evaluation and re-evaluation over several days as drying continues. The amount of time necessary to complete the process varies dramatically. The critical element to the process is that change must be constant. If drying progress is not being made, the restorer must facilitate change in one way or another. The most economical and effective methods are chosen through a constant focus on contamination, damage and cost.

Throughout the process materials will return to normal moisture levels. The restorer must constantly ask "Is it dry?" **Is it done?** during the monitoring process. This means the restorer must know what "dry" is. The materials must be compared against a dry standard (drying goal). As materials return to their dry standard, equipment is reduced. Only after all drying is complete is equipment removed entirely.

It is important that restoration processes and procedures are carefully evaluated to ensure that they are appropriate. Because every job is different, the approach should also be different. Being able to adapt and change to ensure a proper fit between process and structure will translate to significant time and cost savings. The Restorative Drying Process was designed to ensure that each critical and dynamic variable is considered.

Chapter 1

Communication

IN THIS SECTION

INTRODUCTION

The effectiveness of an individual or organization is seriously limited by poor or non-existent communication. With effective communication, a restorer can justify decisions and effectively accomplish job coordination. Poor communication makes these tasks more difficult or even impossible.

The importance of good communication with all materially interested parties including those with an interest in the work being completed cannot be overemphasized. Restoration effectiveness depends on this skill. Poor communication produces a range of undesirable consequences including disorganization, work completed out of correct sequence, additional damage or potential health concerns – just to name a few.

Effective communication requires that the restorer be proficient in a number of communication skills – both verbal and non-verbal. The words that are used, the manner in which they are delivered, the body language adopted, the written documentation and the corporate image presented all play their part in how the restorer's message is conveyed and how it is received by all parties involved.

The restorative drying process disrupts the lives of building occupants. It is not unusual for them to try to alter the drying system to reduce this disruption. These alterations hinder drying progress. Good communication throughout the drying process helps meet this challenge and many others.

FIRST IMPRESSIONS

People tend to develop an impression of someone they meet for the first time very quickly. Studies have shown that people make a decision about someone else within 30 seconds of meeting them, and that they tend to hold on to that first impression for a very long period of time. It is easy to overlook many of the things that influence a first impression. Even prior to the face-to-face meeting, the restorer's uniform, vehicle and posture have established the majority of the client's first impression.

Establishing a positive image is the first step in ensuring that the client will have faith in the restoration efforts that follow. For this reason, a convincing level of professionalism in appearance should be visible to all materially interested parties including:

- Building owners
- Occupants
- Insurance adjusters
- Insurance company representatives
- Subcontractors and even mortgage companies

Items that influence a property owner's first impression include:

- Attitude
- Driving habits
- Appearance of technicians
- Technician's greeting to client
- Certification of company and of technicians
- Appearance of company truck and equipment
- Communication skills – ability to explain what needs to be done and why

Even before the technician arrives to the job site, first impressions begin with the initial phone call.

INITIAL PHONE CALL

Prior to the initial job site visit, a skilled company representative acquires critical information that will better prepare the on-site restorer's response and ensure that a positive first impression is established. Knowing what to ask during the initial client phone call is only half the battle. Knowing *how* to ask takes specialized training and experience. This initial conversation is crucial to obtaining the information required to prepare for the job itself and to establish good rapport with the client.

Communication

First, the restoration contractor needs to determine the category and classification of loss *prior to arrival.* A well-educated and trained representative will know what questions to ask in order to gather this information. For example:

- Instead of: *"What category of water is in your home, ma'am?"*
- A skilled contractor asks specific questions such as: *"What is the origin of the water?/ Where did it come from?" "How long has the structure been wet?" "Where is it wet within the structure? (Where might water have traveled?)"*

The company representative must also be able to address the client's misfortune in a sympathetic way over the phone, assuring them that the most important step in returning their home to its original condition has already been accomplished: calling their restoration company.

There are many pieces of important information that must be collected during this initial phone call. A sample call report form is included in the appendix of this manual. It includes many of the questions the restorer needs to ask. It is necessary to determine the general nature of the damage before making commitments to clients regarding loss mitigation and restoration capabilities. The *S500 Water Damage Restoration Standard and Reference Guide* provides guidelines for information gathering during the initial phone call. Questions include:

- What is the source and condition of the water?
- When did the damage occur?
- Do you own or lease the property (insurance information)?
- What steps have been taken to repair or eliminate the source?
- Which areas are affected (number of rooms, size)?
- How much furniture is present in the affected areas (possible staining, personnel requirements)?

Restorers must be prepared to answer the call 24 hours a day, 7 days a week.

- Has there been previous water damage (source and remediation)?
- Is there electricity and water available (lighting, equipment operation, chemical mixing)?

Following the information gathering, it is important to discuss the extent and category of water, focusing on the health and safety of the client and on mitigating any further loss to the contents and structure.

The initial phone call presents an opportunity to offer guidance to the client on some "do's" and "dont's". Items to encourage the client to *do*:

DO ask clients to:

- ❑ Eliminate the source of water if possible.
- ❑ Turn off circuit breakers if safe to do so.
- ❑ Remove and secure small furniture if possible.
- ❑ Protect carpet from staining using foil (plastic furniture tabs, foam blocks, plastic wrap) beneath furniture legs.
- ❑ Hang draperies and pin up furniture skirts.
- ❑ Remove books, shoes, paper goods, fabrics, potted plants and other items that may stain carpet (or be damaged by water).
- ❑ Remove and secure breakables.
- ❑ Contact a doctor if you have any health concerns.
- ❑ Be cautious of slip hazards.

Documentation begins with the initial phone call.

Suggestions as to what *not* to do are just as important:

- Activities that may put the building owner and/or occupants at risk of electrical shock
- Exposure to toxins or any other physical hazard.
- Activities that may create secondary damage from moisture.

This list will vary depending upon the type of water affecting a structure and the degree of intrusion.

Ensure That The Client DOES NOT:

- ❑ Use vacuum equipment to remove water.
- ❑ Use newspaper on wet surfaces.
- ❑ Walk across wet surfaces any more than necessary.
- ❑ Activate the building **HVAC** system.
- ❑ Adjust indoor temperatures.
- ❑ Enter an area with standing water.
- ❑ Consume any food in the loss area, especially if water was from an unsanitary source.
- ❑ Use personal hygiene items left in the affected area, especially if water was from an unsanitary source.
- ❑ Use home fans to dry wet surfaces.

DOCUMENTATION REQUIRED

There is a legal axiom that the restorative drying industry would do well to follow: *"If it is not documented, it did not happen."* Documentation plays a significant role in communication. The records produced through documentation during the restorative drying process retain job specific information with much greater accuracy and credibility than any other form of communication. High quality and diligent accuracy will ensure that the information sends the message that it is intended to deliver, and will get the invoice paid faster. Incomplete paperwork, sloppy penmanship, inaccurate information or inconclusive data damage client rapport and can lead to incorrect assumptions.

Paperwork begins with the initial telephone call. Upon arrival, documentation continues with the completion of a **Work Authorization Form** (contract) for the services to be provided.

A Work Authorization clearly states the agreement between the building owner and the restoration contractor. This agreement is discussed verbally and the restorer makes written notes as each significant item is covered, to ensure that the materially interested parties fully understand the contract they are entering.

Communication

The Work Agreement is signed before work begins. Using paperwork as a guide, the restorer briefs clients on the course of action. Most building owners and occupants are unaware of the steps necessary to complete the drying and restoration process. Most clients will assume the work will be completed in a rather short period of time. In many cases, this is not true. Realistic expectations need to be set with the client.

A **Customer Briefing/Work Authorization Form** aids in explaining the steps necessary to restore the structure to a pre-loss condition. It is important to set expectations appropriately so that the restorer has the opportunity to be successful. If expectations are set too high, the client will walk away from the experience with a negative impression of the work completed. Restorers must therefore take time to consult with the client and adjust expectations on what needs to be done, what can be done and how long it will take to get it done.

Another important piece of documentation as it pertains to communication is the **Initial Inspection and Loss Assessment Report.** Drying decisions are made based upon critical information obtained during the initial assessment. This information must be gathered and recorded in the job file so that it is available for future review by all parties involved in the restoration project. It is also necessary to justify actions taken by the restorer.

The extent of the water intrusion/migration must be documented. Photos are especially helpful.

Information included in an Initial Inspection and Loss Assessment Report includes:

- Identification and evaluation of health concerns and safety hazards
- Determination of the source of water intrusion
- Determination of the need to protect floor coverings and contents
- Determination of the extent of moisture intrusion
- Determination of the job scope
- Evaluation of flooring materials
- Evaluation of inventories and/or content items
- Evaluation of the HVAC system
- Evaluation of basements and crawlspaces
- Assessment of other structural materials (walls, ceilings, etc.)
- Documentation of **pre-existing conditions** not related to the current loss
- Established drying goals (**dry standards**)

Following the Initial Loss Assessment, daily records of moisture levels in the structure will assist in communicating the importance of each consecutive step in the restorative drying process. This information is recorded in a way that is easily understood – not only by the technician monitoring the job site, but also by all other materially interested parties.

Many templates are available today that make it easy to produce complete, professional documentation including key data and graphical illustrations. This type of documentation is essential in ensuring the reader clearly understands that progress is being made.

It is also important to clearly communicate the predetermined drying goal so that the client understands when to anticipate completion. Restorers build client confidence by ensuring that the finish line is understood and progress is evident and can be tracked.

Before restoration services are complete, materially interested parties review final documentation and the property owner or authorized agent signs an **Acknowledgment of Satisfactory Completion** form. Any remaining questions or concerns by any of the materially interested parties are noted and addressed at this time.

Other documentation may be necessary, depending upon the scope of work. This additional documentation may include:

- Contents and personal property inventories
- Schedule of Loss
- Informed Consent of Product Use forms (if biocides are used)
- Client Refusal of Service Waiver
- Change orders
- Estimates, invoices and bills
- Subcontractor contracts
- Permits
- Contact information for other interested parties
- Detailed work activity log
- Equipment logs
- Lien notices and lien releases

Samples of many of the forms discussed in this chapter are included in the appendix and on the CD.

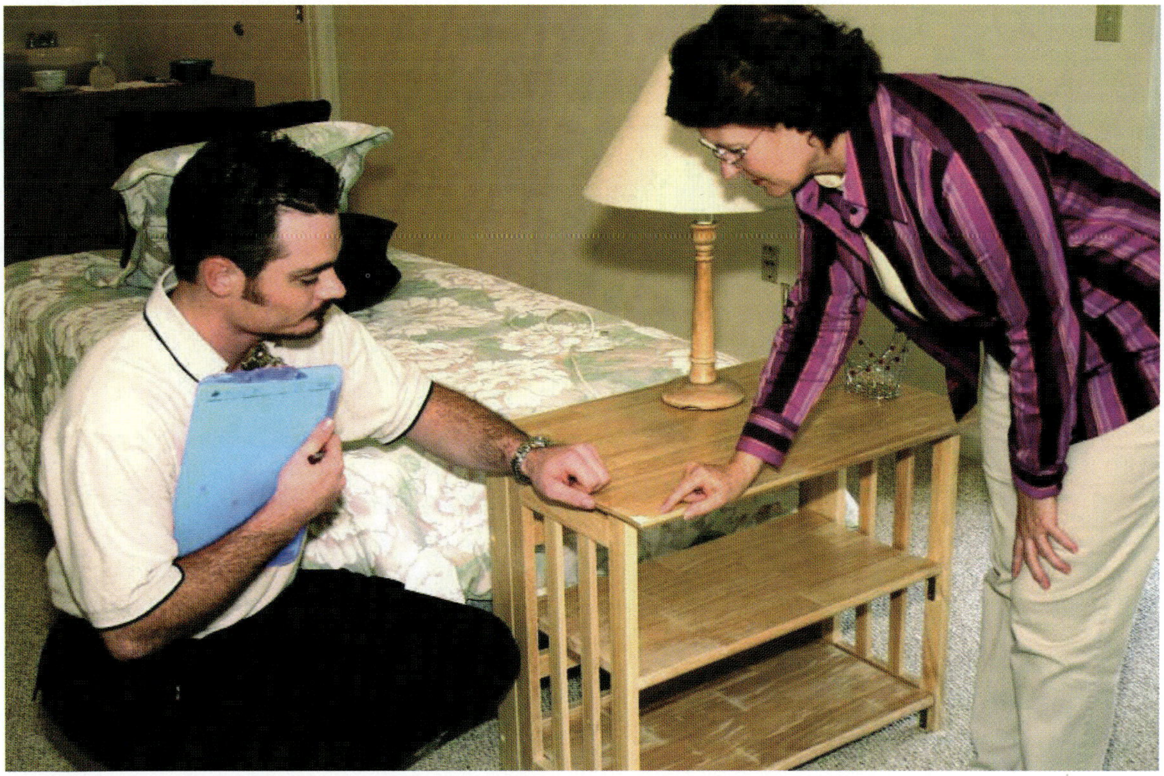

Communication with the client begins with a professional appearance.

Chapter 2

Safety

IN THIS SECTION

INTRODUCTION

The technician's first responsibility when arriving at a water damage site is to *identify and eliminate safety hazards.* Potential safety hazards in water damaged environments include "slip and fall" accidents, electrical shock and falling debris. The restoration contractor is required to provide for the health and safety of workers and occupants during restoration procedures. This responsibility begins with proper employee training so that technicians can quickly recognize safety hazards and effectively communicate each hazard to all parties who could potentially be affected. Site-specific safety information and company procedures must be reviewed with employees when project conditions indicate this need.

CODE OF FEDERAL REGULATION (CFR)

Two codes are important to restoration contractors. These documents can be accessed on-line:

- 29 CFR 1910 - General Industry Standards
- 29 CFR 1926 - Construction Industry Standards

www.osha.gov

Issues covered by these two groups of codes include:

- OSHA General Duty Clause
- Emergency Action and Fire Prevention Plans
- Personal Protection Equipment
- Respiratory Protection Plans
- Asbestos and Lead
- Heat Disorders and Health Effects
- Confined Spaces
- Hazard Communication Plans
- Lock Out/Tag Out Procedures
- Fall Protection
- Noise Exposure Limits

The OSHA General Duty Clause (Sec. 5) states that:
www.osha.gov/dts/shib/shib101003.html

(a) Each employer: (1) shall furnish to each of his employees employment and a place of employment which are free from recognized hazards that are causing or are likely to cause death or serious physical harm to his employees; (2) shall comply with occupational safety and health standards promulgated under this Act.

(b) Each employee shall comply with occupational safety and health standards and all rules, regulations, and orders issued pursuant to this Act which are applicable to his own actions and conduct.

Preparing a company **Safety Plan** that addresses each of the items listed above will assist in minimizing potential injury while performing restoration services. Documenting this plan and frequent employee training will also assist in complying with federal regulations.

As stated on the OSHA website:
www.osha.gov/dts/shib/shib101003.html

The Occupational Safety and Health Act requires employers to comply with hazard-specific safety and health standards as issued and enforced by either the Federal Occupational Safety and Health Administration (OSHA), or an OSHA-approved State Plan. In addition, Section 5(a)(1), the General Duty Clause, requires employers to provide their employees with a workplace free from recognized hazards likely to cause death or serious physical harm.

Employers can be cited for violating the General Duty Clause if there is such a recognized hazard and they do not take reasonable steps to prevent or abate the hazard. However, failure to implement these guidelines is not, in itself, a violation of the General Duty Clause. Citations can only be based on standards, regulations and the General Duty Clause.

The most common safety hazard in a wet structure is the risk of slipping and falling.

COMMON HAZARDS

The most common safety hazards associated with water damaged structures result in "slip and fall" accidents. Although slippery floors are the most widespread hazard in water damaged structures, they are not necessarily the most dangerous hazard – and by no means do they represent the only one.

Additional hazards include (adapted from *S500 Standard and Reference Guide For Water Damage Restoration):*

- Compromised structural integrity
- Contaminated water
- Shock hazards and gas leaks
- Stored chemicals
- Damaged construction components and materials
- Household chemicals
- Commercial or industrial chemical waste
- Microbial contamination
- Lead and Asbestos
- Heat disorders
- Confined spaces (crawlspaces and attics)

TIP

A safety plan should be documented and discussed with personnel regularly. A monthly safety meeting can be used to cover important company safety policies and to review recent hazards encountered in the field.

Safety

To increase safety and reduce company liability when performing residential water damage restoration, contractors should leave caution signs, use safety certified equipment (i.e. **UL**, **CSA** and **CE**), use ground fault circuit interrupters (**GFCI**), and warn clients of "slip and fall" hazards and any other possible threats to their safety and health.

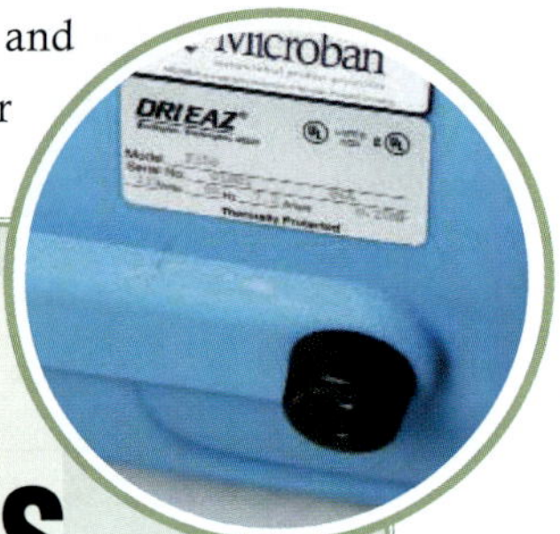

Restorers should ensure this label appears on all drying equipment (or similar labels, such as CE, CSA).

Once a hazard has been identified, the first line of defense is to *eliminate and communicate:*

1. *Eliminate* as many potential hazards as possible.
2. *Communicate* any remaining potential safety concerns to all involved parties.

Once identified, documented and communicated, any further steps appropriate for eliminating the hazard are implemented.

With each step performed during restoration, the jobsite must be re-evaluated for additional task-related safety hazards. Additional hazards will arise, particularly during the demolition process.

As OSHA reports, occupational injuries occur for a variety of reasons:
www.osha.gov

> *OSHA has determined that workers involved in a wide range of occupations are exposed to a significant risk of death or injury from being struck by various objects in the workplace. OSHA's incident data indicate that a significant portion of all work related injuries and fatalities involve workers being struck in the eyes, head, face, hand and/or feet by foreign objects. For example, it has been estimated that as many as 2,500 eye injuries occur in the workplace every working day.*

OTHER HAZARDS: LEAD AND ASBESTOS

It is important not to overlook two other hazards that may be present in a water-damaged environment: *lead and asbestos.* Specific laws and regulations apply especially to the handling of asbestos, and they must be followed. (The full CFR can be found on-line at www.osha.gov.)

According to OSHA, **asbestos** must be handled and removed only by qualified personnel; failure to comply with this regulation can result in substantial fines and potential lawsuits. This regulation must be observed whenever handling asbestos containing materials (ACM) – defined as any material containing more than 1% asbestos by weight.

As stated in the Code of Regulation (29 CFR 1926.1101):

> *1926.1101(e)(6) Competent Persons. The employer shall ensure that all asbestos work performed within regulated areas is supervised by a competent person, as defined in paragraph (b) of this section. The duties of the competent person are set out in paragraph (o) of this section.*

> *1926.1101(b) Competent person means, in addition to the definition in 29 CFR 1926.32 (f), one who is capable of identifying existing asbestos hazards in the workplace*

TIP

Structures are at a higher risk for ACM (asbestos containing material) if they were built prior to 1981. Suspect materials need to be inspected and tested prior to demolition.

and selecting the appropriate control strategy for asbestos exposure, who has the authority to take prompt corrective measures to eliminate them, as specified in 29 CFR *1926.32(f); in addition, for Class I and Class II work who is specially trained in a training course which meets the criteria of EPA's Model Accreditation Plan (40 CFR 763) for supervisor, or its equivalent and, for Class III and Class IV work, who is trained in a manner consistent with EPA requirements for training of local education agency maintenance and custodial staff as set forth at 40 CFR 763.92 (a)(2).*

Asbestos is also addressed in 29 CFR 1910.1001 as it pertains to occupational exposure, whereas the CFR 1926.1101 addresses asbestos exposure in construction work.

As a restorer, it is important to understand where asbestos containing materials might be found in the structure and how to recognize that asbestos may be present. For example: 9" x 9" (23 x 23 cm) flooring tile can be perceived as vinyl composite tile (VCT), but may be vinyl asbestos tile (VAT). Pipe insulation, wall insulation, several glues and adhesives and some ceiling textures are all examples of other potential asbestos containing materials.

Lead is also addressed in two significant codes of federal regulation. In 29 CFR 1026.62, OSHA addresses lead in the construction work environment. In 29 CFR 1910.1025(a)(1), OSHA addresses lead as it applies to all occupational exposure.

As with asbestos, if lead is identified on the job site, all demolition work must stop. A licensed lead abatement contractor must be contacted if such material is disturbed.

The following are excerpts from 29 CFR 1926.62 - Lead:
www.osha.gov

1926.62(a) This section applies to all construction work where an employee may be occupationally exposed to lead. All construction work excluded from coverage in the general industry standard for lead by 29 CFR 1910.1025(a)(2) is covered by this standard. Construction work is defined as work for construction, alteration and/or repair, including painting and decorating.

1926.62(e)(1) Engineering and work practice controls. The employer shall implement engineering and work practice controls, including administrative controls, to reduce and maintain employee exposure to lead to or below the permissible exposure limit to the extent that such controls are feasible. Wherever all feasible engineering and work practices controls that can be instituted are not sufficient to reduce employee exposure to or below the permissible exposure limit prescribed in paragraph (c) of this section, the employer shall nonetheless use them to reduce employee exposure to the lowest feasible level and shall supplement them by the use of respiratory protection that complies with the requirements of paragraph (f) of this section.

PERSONAL PROTECTION EQUIPMENT (PPE)

The first line of defense against potential safety hazards is awareness. Due to the nature of water-damaged structures, it is not always possible to eliminate all health and safety concerns. For this reason, personnel must use appropriate personal protection equipment (PPE).

It is important to ensure a proper level of protection when considering PPE. Risks should be eliminated, when possible, through other means, negating the need for PPE, or risks can be minimized to reduce the level of personal protection needed. A close evaluation of the potential risks is necessary in order to identify which type of PPE would be most appropriate.

Types of PPE

There are many forms of PPE that may be required to perform the tasks associated with restorative drying. OSHA's introduction of 29 CFR 1910.132 reads:
www.osha.gov

Protective equipment, including personal protective equipment for eyes, face, head, and extremities, protective clothing, respiratory devices, and protective shields and barriers, shall be provided, used, and maintained in a sanitary and reliable condition wherever it is necessary by reason of hazards of processes or environment, chemical hazards, radiological hazards, or mechanical irritants encountered in a manner capable of causing injury or impairment in the function of any part of the body through absorption, inhalation or physical contact.

Safety

CHAPTER 2

To address the many hazards associated with restorative drying work, contractors need to have ready access to several items. These items include:

- Chemical resistant gloves
- Splash goggles
- Appropriate respirator
- Hard hat
- Rubber boots with hardened shank/last
- Protective clothing

Respirators

An estimated 5 million workers are required to wear respirators in 1.3 million workplaces throughout the United States. Respirators protect workers against harmful dusts, fogs, smokes, mists, gases, vapors and sprays. These hazards can cause cancer, lung impairment, other diseases or even death.

An alternative to a respirator may be ventilation. In many cases, respiratory protection needs can at least be minimized when an environment with respirable hazards is ventilated.

The use of PPE requires proper training and established company procedures, especially when there is a requirement to use respiratory protection on the job site. Before commencing work as a restorer, it is essential that a medical evaluation be performed by a qualified physician or when not possible, be completed online. This medical evaluation ensures that the restorer is medically fit to work in this type of equipment. It also provides the restoration employee with a baseline of respiratory health at the beginning of employment.

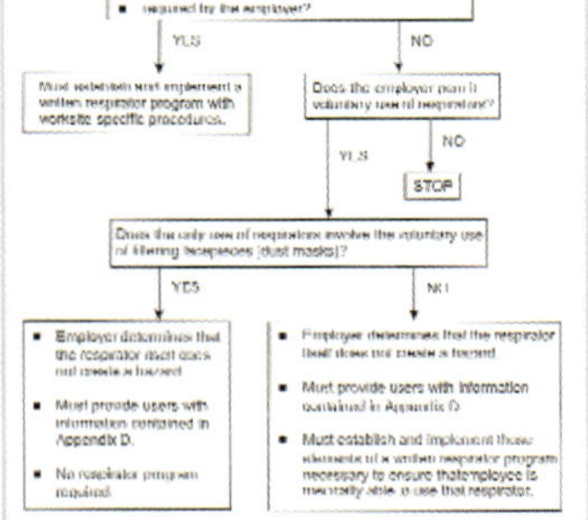

www.osha.gov/SLTC/etools/respiratory/oshafiles/require.html

Use of respiratory equipment must conform to OSHA 29 CFR 1910.134(c), which states that the employer must provide training, fit testing (ensuring respirator fits properly) and medical evaluations. OSHA provides a flowchart for evaluating whether it is necessary for employers to provide respirator training and fit testing, which has been included in the appendix of this book. When using respiratory protection, it is important to choose the appropriate respirator for the job. This selection is primarily based on the hazards to which the worker will be exposed to during the course of the job. However, the particular respirator that will work best also depends upon the specific needs of the individual wearing the respirator:

- The nature of the hazard
- The location of the hazardous area
- The employee's health
- The work activity being undertaken
- Characteristics, capabilities and limitations of the respirator itself

Appropriate use of PPE may require several components and proper training on the use of each.

Respirator types vary, but include the following major classes:

Class 1: Air Purifying Device. This class of respirator is equipped with an air-purifying filter, cartridge or canister that removes specific air contaminants by passing ambient air through an air-purifying element. The air-purifying device (the most common type of respirator used in the restoration industry today) cleanses the contaminated atmosphere. This type of respirator is limited in its use to those environments where the air contaminant level is within the specified concentration limitation of the device. It is necessary that sufficient oxygen exists in the area, since the respirator does not supply oxygen. There are various types of air-purifying device technology, including:

- Mechanical-filter cartridge
- Chemical cartridge
- Combination mechanical-filter/chemical cartridge
- Gas masks
- Powered air-purifying respirators

Class 2: Atmosphere-Supplying Respirator or Supplied-Air Respirator (SAR). This class of respirator supplies the user with air from a source other than the ambient atmosphere, and includes supplied-air respirators (SARs) and self-contained breathing apparatus (SCBA) units. With SARs, air is provided through a supply hose connected to the wearer's face piece or enclosure. Three basic classes of air-line respirators exist: continuous flow, demand-flow and pressure-demand flow.

Class 3: Combination Air-Purifying and Atmosphere Supplying Device. This class of respirator combines an air-line respirator with an auxiliary air-purifying attachment, which provides protection in the event the air supply fails. The filtering mode is typically used for escape purposes only. Because they work on positive pressure and have escape provisions, these respirators are highly recommended for use in asbestos type work.

Respirator Cartridges. Class 1 and 2 respirators include air purifying filters. These filters are contained in disposable cartridges. The filters remove particulate or vapors from the air. Several cartridge types are available. They are color coded to indicate what particulate or vapor they can remove from air:

- Magenta – P-100 (HEPA) for particulate
- Black – organic vapor
- Yellow – organic vapor/acid gas
- White – organic vapor/acid gasses/formaldehyde
- Green – ammonia

Color coding may vary from country to country.

The respirator cartridges most typically used in a restorative drying environment are those for particulate (magenta) and for organic vapor (black). Stacked or combination cartridges are also available.

Additional Safety Equipment

It is necessary to make additional safety equipment available in order to be prepared for the number of health and safety concerns that can arise. Some of the additional safety equipment to keep on hand includes:

- Fire extinguisher
- First aid kit
- Chemical Material Safety Data Sheet (MSDS)
- Hazard warning signs
- Electrical circuit testers

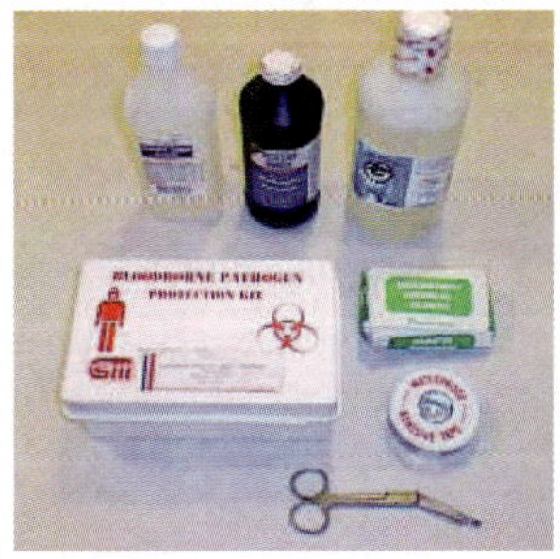

First Aid Kits should be readily available at the worksite.

IMMUNIZATIONS

As a minimum medical requirement, technicians performing water damage restoration services need medical consultation for appropriate immunizations (shots). Immunizations required vary depending upon the scope of work, previous immunization history, previous exposure and current availability of treatments.

The physician needs to be made aware of the environments in which the individual will be working, then asked for advice on what immunizations and boosters are needed to provide adequate protection. It is important to understand that common restoration environments involve direct contact with a range of infectious organisms. Of particular concern are Hepatitis A and B and Tetanus. Exposure risk is reduced through proper use of PPE and appropriate immunization.

Additionally, restorers with immunosuppressing medical conditions need to consult with a physician to fully understand the increased risks of exposure when working in an environment affected by water intrusion.

SAFETY PROCEDURES

Confined Space Entry

A confined or enclosed space is defined as any space that:

- Is configured so that an employee can enter it;
- Has limited means of entry and exit;
- Has the potential for containing atmospheric hazards; and
- Is not designed or intended for continuous occupancy.

Confined spaces that fit the description of "permit required" are those that contain:

- Atmospheric hazards which are or potentially can be asphyxiating, toxic, flammable or explosive;
- Engulfment hazards which exist when someone is trapped or enveloped by a liquid or dry bulk material such as grains, soil or powdered cement;
- Configuration hazards wherein the size or shape of the space can trap a worker or make escape or rescue difficult; or
- Any other recognized serious safety health hazard.

Some environments in a water loss fit the definition of a confined space (i.e., a crawlspace or an attic). Entry into such a space requires specific procedures, training and, in some cases, special permits. Depending upon circumstances, the procedures require the use of:

- Attendant – an individual stationed outside one or more permit spaces who monitors the authorized entrants and who performs all other attendant's duties as assigned in the employer's permit space program.
- Retrieval system – the equipment (including a retrieval line, chest or full-body harness, wristlets, if appropriate, and a lifting device or anchor) used for non-entry rescue of persons from permit spaces.
- Testing – planning and implementing tests in order to identify and evaluate hazards that may confront entrants of a permit space.
- Specialty PPE – the identification and wearing of specialty PPE appropriate for the confined space, which may include supplied oxygen.

Some environments will necessitate extensive use of PPE and may require specific procedures to ensure the safety of workers.

OSHA regulations that address confined spaces are found in 29 CFR 1910.146 (for permit required confined spaces) and 29 CFR 1926.21 (safety and health regulations for construction – safety training and education), and further information is available from the American National Standard Institute in ANSI Z117.1-1989, Safety Requirements for Confined Spaces. Before entering any space that is not designed to be an occupied space, technicians need to be aware of the information outlined in the ANSI standard and OSHA CFR.

Lockout – Tagout

When working in spaces where the unexpected start up of machines and equipment or the release of stored energy could cause injury or death, restoration employees must use appropriate measures to identify, disable and lockout-tagout the potential hazard. Equipment capable of causing injury in a residential application may include HVAC equipment, electrical systems or equipment containing combustible fuel. When working in an industrial or commercial environment, the list of potential equipment-related hazards varies greatly.

Stored energy sources should be locked-out with a tag if they can cause injury to workers.

A thorough evaluation of the surroundings and contents, along with a working knowledge of the normal use and application of a particular space or environment, will assist in determining whether such hazards are present.

Where these hazards are present, elimination of the source of the hazard can include a lockout-tagout program. OSHA 29 CFR 1910.147 addresses the procedures and policies necessary to properly disable and prevent unexpected startup or energy release from these types of hazards.

Proper lockout-tagout procedures may include shutting down power to the source of the hazard, labeling the item and using physical locks to ensure that the power is not restored.

Safe Work Practices

In addition to the other safety practices outlined in this chapter, there are a number of general safety protocols to follow when working in a water damaged structure. The general practices outlined below are designed to help prevent exposure to hazards in a restorative drying environment:

- ☑ Before entering any new area of a structure, stop to observe and identify any potential safety or health hazards that may be present.
- ☑ Document and report any observed safety hazard immediately.
- ☑ Avoid eating, drinking, smoking, chewing tobacco and the application of personal hygiene items while on the job site.
- ☑ Wash hands and face after removing any PPE.
- ☑ Ensure all PPE is properly cleaned and maintained after each use.
- ☑ Shower after each work day.
- ☑ Launder all clothing after each work day.

DOCUMENTATION

Several areas of health and safety protection require proper documentation. This section will discuss some of the key documentation necessary in the restorative drying industry as it pertains to health and safety.

Respiratory Protection Plans

The use of respiratory protection requires the use of a written **respiratory protection plan.** OSHA 29 CFR 1910.134 outlines respiratory protection codes as follows:

> *Respirators shall be provided by the employer when such equipment is necessary to protect the health of the employee. The employer shall provide the respirators which are applicable and suitable for the purpose intended. The employer shall be responsible for the establishment and maintenance of a respiratory protection program which shall include the requirements outlined in paragraph (c) of this section.*

OSHA provides all of the information necessary in order to develop and maintain a written respirator protection plan in 29 CFR 1910.134, with sections including:

- *1910.134 - Respiratory Protection Standard*
- *1910.134 - App. A - Fit Testing Procedures (Mandatory)*
- *1910.134 - App. B-1 - User Seal Check Procedures (Mandatory)*
- *1910.134 - App. B-2 - Respirator Cleaning Procedures (Mandatory)*
- *1910.134 - App. C - OSHA Respirator Medical Evaluation Questionnaire (Mandatory)*
- *1910.134 - App. D - Information for Employees Using Respirators When Not Required Under the Standard (Mandatory)*
- *Interpretations for 1910.134*
- *Respiratory Protection - Safety and Health Topics*

Hazard Communication (HAZCOM)

Whenever hazardous chemicals are present in the work environment, HAZCOM programs must be documented. HAZCOM programs include the use of training, labeling systems and forms of warning. The OSHA 20 CFR 1910.1200 addresses the creation of such programs.

www.osha.gov

OSHA's introduction of 29 CFR 1910.1200 reads:

> *The purpose of this section is to ensure that the hazards of all chemicals produced or imported are evaluated, and that information concerning their hazards is transmitted to employers and employees.*
> *This transmittal of information is to be accomplished by means of comprehensive hazard communication programs, which are to include container labeling and other forms of warning, material safety data sheets and employee training.*

Blood-Borne Pathogen Plan

Workers in many different occupations are at risk of exposure to bloodborne pathogens such as Hepatitis B, Hepatitis C and HIV/AIDS. First aid team members, housekeeping personnel in some settings, water damage restoration technicians and nurses are examples of workers who are at risk of exposure. In 1991, OSHA issued the Bloodborne Pathogens Standard to protect workers from this risk.

Restorers should ensure a proper fit whenever using respiratory protection such as the full face respirator shown here.

Bloodborne pathogens are microorganisms present in the blood which can cause disease in humans. These pathogens include the Hepatitis B virus (HBV) and the Human Immunodeficiency Virus (HIV). It is very possible that restorers will come into contact with bloodborne pathogens during day-to-day restoration activities, primarily through the remediation of sewer back-ups.

Restorers need to understand the seriousness of these pathogens, know what PPE to employ, and what procedures and standards must be followed. When documented, this information becomes a bloodborne pathogen plan for their business. Just as important as developing the plan is implementing the plan, which includes training for workers on the protocols and procedures to be used.

Emergency Action Plans

Yet another area discussed in multiple OSHA CFR documents is the use and management of

Information regarding chemicals present in the work area should be clearly posted and readily available at all times.

Emergency Action Plans. Organizations with more than 10 employees must have written plans, while organizations with fewer than 10 employees may orally communicate plans to staff.

Emergency action plans must include procedures for: fire reporting, emergency evacuation, operation of critical equipment during evacuation, how to account for staff, rescue operations, and contact information for those who can provide additional information. An excerpt from 29 CFR 1910 outlines the required scope for an Emergency Action Plan:

www.osha.gov

1910.38(b) Written and oral emergency action plans. An emergency action plan must be in writing, kept in the workplace, and available to employees for review. However, an employer with 10 or fewer employees may communicate the plan orally to employees.

An emergency action plan must include at a minimum:

- *1910.38(c)(1) Procedures for reporting a fire or other emergency;*
- *1910.38(c)(2) Procedures for emergency evacuation, including type of evacuation and exit route assignments;*
- *1910.38(c)(3) Procedures to be followed by employees who remain to operate critical plant operations before they evacuate;*
- *1910.38(c)(4) Procedures to account for all employees after evacuation;*
- *1910.38(c)(5) Procedures to be followed by employees performing rescue or medical duties; and*
- *1910.38(c)(6) The name or job title of every employee who may be contacted by employees who need more information about the plan or an explanation of their duties under the plan.*

ORGANIZATIONS AND STANDARDS

There are several sources for additional information regarding worker and occupant safety. Many of these sources are not specific to the restorative drying industry, but are specific to several of the job-related tasks performed by restoration technicians.

Standards and organizations that are specific to restorative drying typically refer, in large part, to these other non-specific sources. Because many of these sources are federal, much of the content is required practice. It is therefore necessary that restorative drying contractors become familiar with these sources.

The list below contains some of the more relevant organizations and standards restorers need to be familiar with:

www.osha.gov

- **OSHA** - Occupational Safety and Health Administration
  - 29 CFR 1910
  - 29 CFR 1926
- **ANSI** - American National Standards Institute
  - ANSI Z117.1-1989
- **EPA** - Environmental Protection Agency
  - FIFRA - Federal Insecticide, Fungicide and Rodenticide Act
- **IICRC** - Institute of Inspection, Cleaning and Restoration Certification
  - **S500** - Standard and Reference Guide for Professional Water Damage Restoration
  - **S520** - Standard and Reference Guide for Professional Mold Remediation

This list should be evaluated for its application through varying state, provincial and other government laws and regulations. It is also necessary to check with each of these sources frequently in order to stay current with changes to standards, codes and regulation.

Chapter 3

Inspection Tools and Equipment

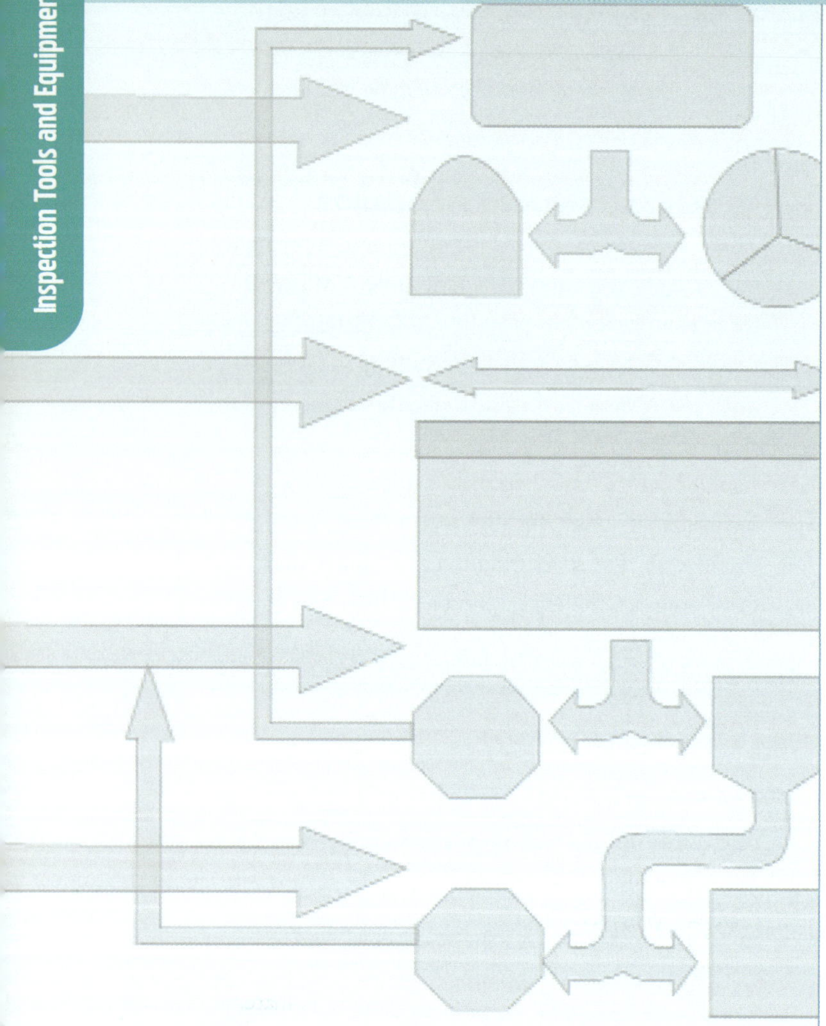

INTRODUCTION

The success of each and every decision made during the restorative drying process depends on the information upon which the decision was based. The process of restorative drying involves a continuous cycle of information gathering, interpretation, decision making, validation and adjustment.

Unique variables exist with every structure, water source and client. It is for this reason that assumption is the root of all failure. This chapter will outline and explore the key tools necessary to perform a thorough inspection and accurate monitoring.

THE FOUR "KNOWS" OF DRYING

There are four questions that summarize the information that must be obtained, documented and evaluated throughout the restorative drying process. When left unanswered, these questions are the source of the most common deficiencies in inspection and documentation. The four key questions, or "knows," of drying are:

1 What's wet?
2. How wet is it?
3. Is it drying?
4. Is it done?

Inspection, monitoring and documentation answer these four questions. A gap in this process will result in improper drying, additional water damage, and compromised structural cleanliness and integrity. Proper instruments and tools are a key ingredient to successful completion of the drying process.

Developments in technology, science, education and practice have all changed the landscape of the restorative drying industry. In particular, instrumentation technology has greatly advanced, offering more and better features for evaluating "What's wet" and "How wet it is." Ultimately, this has led to an increase in liability

IN THIS SECTION

for those who have not kept up to date with current instrumentation technologies and changes in industry standards and procedures. The long-term effects of improper drying have become an area of interest to many attorneys across North America.

Accuracy and thoroughness are accompanied by another important aspect of inspection: *frequency*. Most restorative drying projects need to be evaluated on a daily basis, as every project is unique and presents its own challenges and issues. Some variation in frequency will exist from one project to the next, and how often a project is monitored is based upon multiple factors. Of these the most important are the severity of the loss and the potential for secondary damage to the contents and structure.

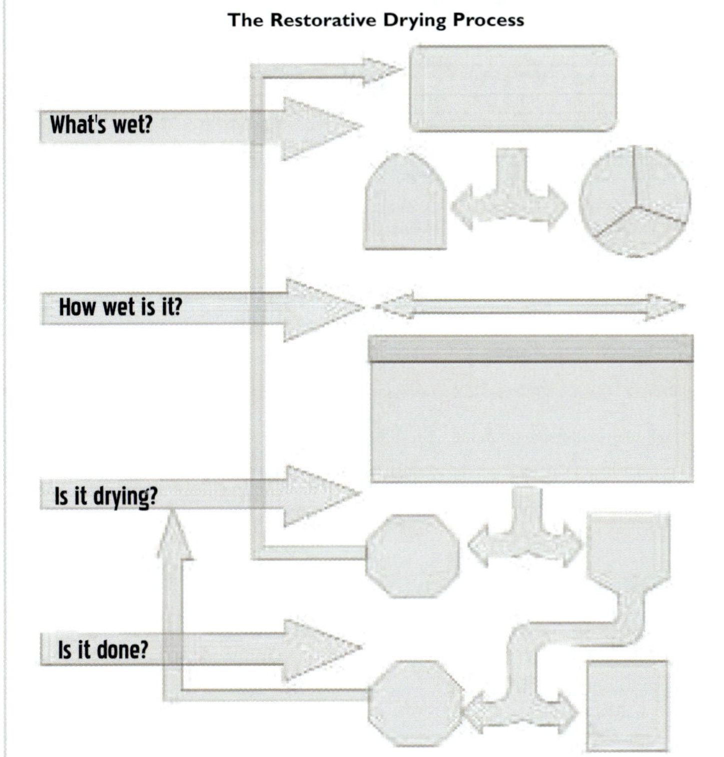

The four "knows" of drying are critical components in solving structural drying problems.

DEFINITIONS AND "TERMS OF ART"

Before exploring inspection and monitoring practices, it is important to understand the terms commonly used when discussing moisture levels, instruments, and documentation.

Percentage of Moisture Content (%MC)

Percentage of moisture content is defined as the amount of moisture in a material compared to its oven-dried weight. Properly measuring the percentage of moisture content in any material requires weighing a material, baking it dry in an oven, and then weighing it again. The weight after the material has been oven dried is then subtracted from the weight of the material before the material was oven dried – giving you the "weight lost" during the oven drying process. The "weight lost" during the drying process is then divided by the oven-dried weight (weight of the material after it has been oven dried). The resulting percentage "was" the material's %MC.

Formula for Percentage Moisture Content:

$$\%MC = W_1 / W_2$$

Where:

%MC is Percentage Moisture Content of the material before oven drying

W_1 is the weight of the water removed during the oven drying process

W_2 is the weight of the material after oven drying

One technology used by meter manufacturers to estimate the material's %MC is electrical resistance (measured in Ohms). This process involves weighing a material when it is wet, and also recording the amount of resistance (Ohms) using an electric meter. The material is then oven dried. After oven drying, the material is again weighed. The oven-dried weight is subtracted from the original wet weight. The resulting weight loss is then divided by the oven-dried weight to determine the actual %MC.

The manufacturer repeats this process several times (using the same material at different moisture levels),

then builds a graph showing the amount of electrical resistance equal to several corresponding %MC. The meter is then set to this data, thereby translating resistance (Ohms) into %MC. A major limitation of this technology is that the meter is then only capable of measuring accurate %MC in the exact material initially used by the manufacturer to set its values.

A common misconception is that meters always measure in %MC. In fact, they rarely, if ever, measure %MC accurately. A number of factors cause the reading to deviate from the actual %MC. These factors include:

Temperature – Changes in material temperature will cause readings to differ from the true moisture content.

Material Variations – Many materials vary in their makeup. Variance in makeup will alter the material's natural resistance.

Meter Calibration – Meters require periodic calibration as the result of normal use.

For these reasons, the term %MC will, in many cases, be an inaccurate term to use when discussing moisture levels found using standard instruments on common building materials. More accurate terms can be used and are discussed in the following sections of this chapter.

Wood Moisture Equivalency *(WME)*

Wood moisture equivalency is a term used to describe readings obtained using wood calibrated moisture meters on non-wood materials. Because most meters are designed with wood moisture content as their intended medium, WME will normally be the most appropriate and consistent term to use when discussing and documenting moisture findings.

WME readings bear no significant value unless compared to some other control sample or benchmark. They are simply a relative reading, where lower numbers represent less potential moisture than higher readings. It is important to consider all factors that might alter the reading of a moisture meter when evaluating the numerical values of an instrument and any changes to that value over time as drying is monitored.

Dry Standard

Establishing a **dry standard** is the best way to set restorative drying goals. A dry standard is established by obtaining a moisture reading using an unaffected material. Regardless of the material type the meter was intended for, unaffected materials will yield a relative reading that can then be used as a control to which suspect materials can be compared.

When establishing dry standard readings, it is essential to document not only the relative value from the unaffected material, but also the meter brand and model, time of year, type of indoor environment and any appropriate material characteristic that may influence the reading obtained. Most critical of these are the meter brand and model used and the specific material being measured.

Restoration contractors set an effective dry standard for use in structural drying by obtaining this data from several materials in several unaffected environments. This then yields a range of normal moisture levels for building materials in the specific geographical region. This record then becomes a reference to evaluate suspect materials in a wet environment. When it is necessary to establish a dry standard for a material within the same structure being evaluated following a water loss, it is important to consider indoor temperature and humidity. Abnormally high humidity, especially, will greatly alter the amount of moisture in materials that are not directly affected by water.

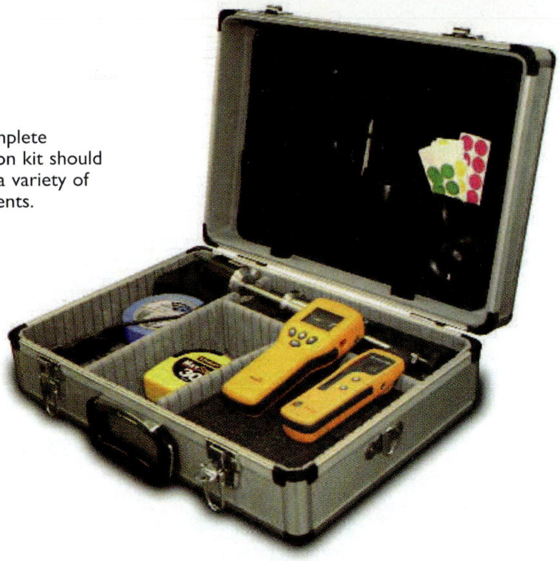

The complete inspection kit should include a variety of instruments.

INSPECTION AND MONITORING TOOLS

There are several tools available for inspection and monitoring. These tools include instruments for detecting moisture, imaging equipment for confined space or cavity inspection, various instruments for temperature and humidity measurements, and devices for measuring the size and volume within the affected areas of a structure. Other tools and devices are also used for measurement in the inspection and monitoring process to assist in the evaluation of environmental and structural conditions.

Moisture Detection Instruments

Moisture Sensor. The **moisture sensor** is the most basic of moisture detection instruments. It is designed to detect the presence of abnormally high levels of moisture in carpet and pad (underlay), indicating "What is wet" not "How wet it is."

A moisture sensor works by using an electrical circuit to test the conductivity of a material through two short, non-insulated, tapered pins that are inserted into the material. Materials with higher levels of moisture conduct electricity better than those that are drier. Thus, where more moisture exists, the instrument has a better connection between the two pins.

Most moisture sensors will indicate the presence of abnormal moisture through an audible beep that increases in frequency if moisture levels are higher, and through a visible LED somewhere on the handle that blinks with a higher frequency when greater amounts of water are present.

Although the amount of moisture does influence the frequency and/or intensity of the instrument's response, no numerical value is given by the device, preventing the technician from evaluating progress. This instrument is therefore a sensor and not a meter. It is used to evaluate water migration through carpet and cushion (underlay) to assist in establishing the size of an affected area.

It is a misuse of the moisture sensor to measure moisture in other materials present in a water-damaged structure, and to evaluate the condition of materials after drying has begun. Because the instrument has been designed only for high levels of moisture in carpet and pad (underlay), it is possible that the meter will give an unaffected response even if materials remain unacceptably wet. Moisture sensors can also supply false indications of moisture migration if high levels of urine are present in the carpet, pad (underlay) or subfloor.

Invasive Moisture Meter (Penetrating). The **invasive moisture meter** operates on a principle that is very similar to the moisture sensor described above. Electrical circuits test the conductivity of the suspect

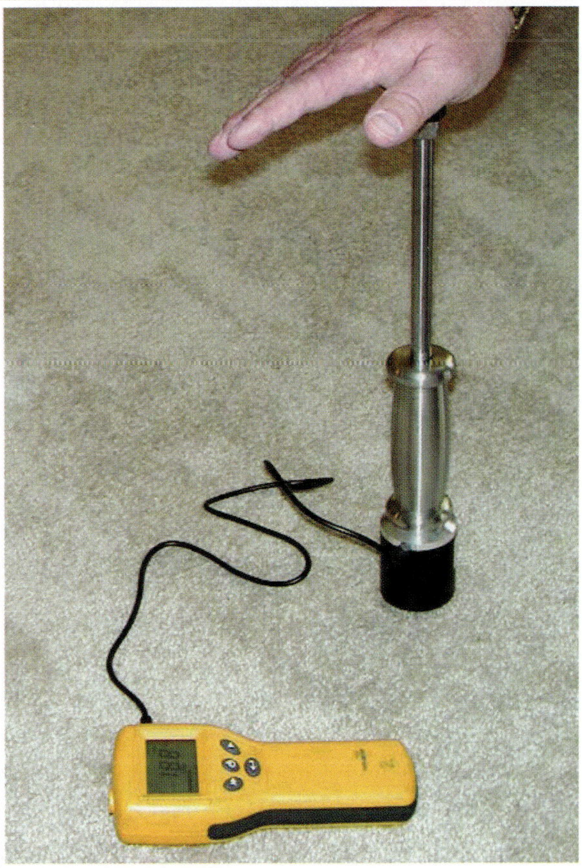

Invasive probes allow for moisture measurement through multiple layers, such as the wooden subfloors beneath this carpet and cushion (underlay).

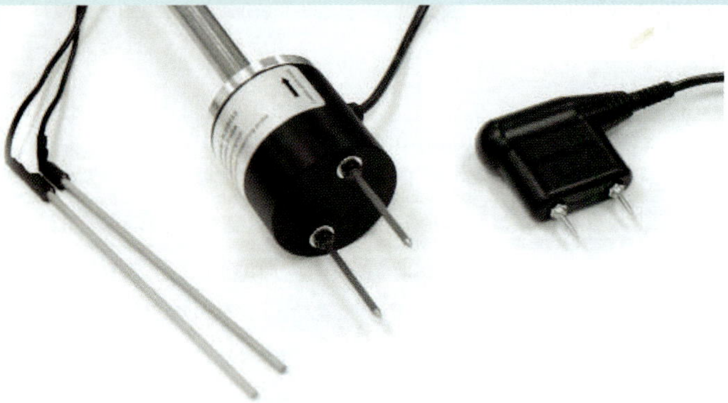

material, and higher degrees of moisture result in a response from the instrument that indicates the target is "wetter."

The primary difference between invasive moisture meters and moisture sensors is the ability to quantify the moisture readings. Moisture meters will quantify the moisture, on either a relative or absolute scale. This numerical value can then be documented and used to monitor progress. For this reason, the amount of moisture absorbed by structural materials is measured most accurately with a moisture meter, not a moisture sensor.

Invasive moisture meters offer a variety of accessories and attachments designed for various building materials and construction methods. These attachments generally include pin (probe) variations in length, diameter, tip style and electrical shielding (insulation) along their shaft. Depending upon the intended medium to be inspected, one probe type may offer better results than another.

Following is a list of the various attachment types and some of their most appropriate applications.

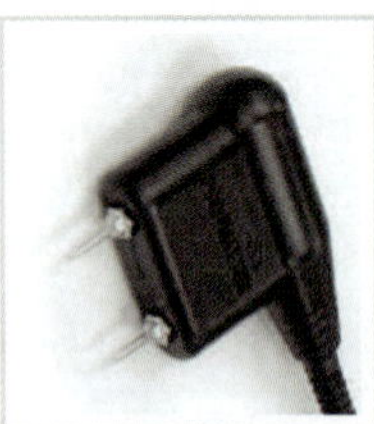

Short (1/2" or 1 cm) tapered, non-insulated pins – These pins are helpful for evaluating materials such as baseboard or, initially, structural materials in the crawlspace. Tapered pins are also very useful in monitoring through thin surface materials such as fabric on upholstered furniture. This pin type will leave the smallest hole behind. In many cases, the invasive moisture meter is manufactured with this pin type permanently attached to the meter housing.

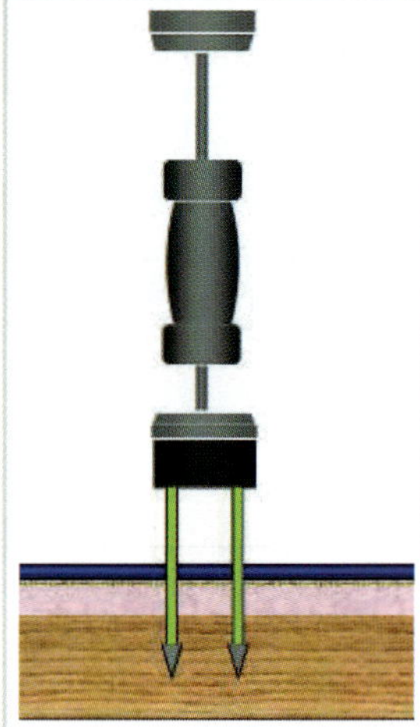

Two-inch insulated pins on a hammer-probe attachment drive into the wood beneath a carpet and pad (underlay).

Medium (2" or 5 cm) straight, insulated pins – Usually associated with a hammer probe, these pins are used to monitor structural wood and wood flooring. The pins' insulation provides isolation as the probe is hammered into varying depths of the material, enabling accurate, single point data at multiple depths. Because the amount of moisture through a material will vary greatly, insulated pins are the most accurate for evaluating actual moisture levels through materials such as wooden subflooring, wood flooring, structural plates, wall studs, and other similar materials. When using a hammer probe to monitor strip hardwood flooring, *the pins are inserted into the dark winter wood* for easier repair and in the center of the board to avoid splitting the strip flooring.

Long (6"- 12" or 15-30 cm) straight, insulated pins – Although the description is similar to the medium pins described above, longer pins are not used on wooden materials because these pins lack the strength necessary to be driven into wood. Long pins are used to evaluate moisture levels behind and beneath finishing surfaces such as gypsum wallboard and cabinetry. Holes are pre-drilled through dense surfaces, or forced through softer materials. This pin type is especially effective when inspecting moisture levels in wall insulation. Long pins can also be used to measure surface moisture on subflooring beneath cabinetry, gypsum wallboard behind cabinetry, and wall plates behind baseboards.

NOTE regarding insulation of the pins – It is important to check the insulation on these pins frequently. As the pins are driven into structural material such as plywood and hardwoods, the insulation will begin to wear. To test the integrity of the insulation, simply run fingers up and down both metal pins, on each side of the pins. If the meter displays elevated numbers, the insulation has been compromised and the pin(s) must be replaced. Failure to replace the pin(s) will lead to inaccurate moisture content readings in the material being monitored.

Non-Invasive Moisture Meter (non-penetrating).

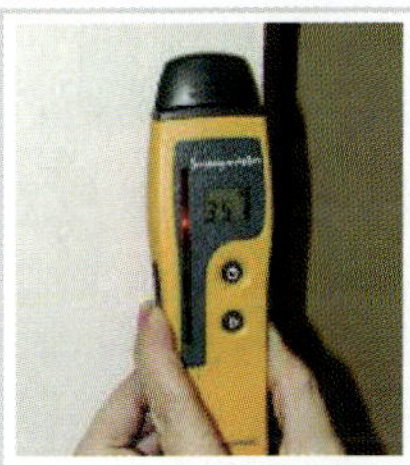

Non-invasive moisture meters use radio frequency signals and conductive pads to measure either impedance or capacitance though a sample of the suspect material. Higher levels of moisture will result in higher transfer of the signal through the material, and therefore will result in higher readings displayed by the instrument. Non-invasive moisture meters sample a much larger area of the material and can survey samples at a much faster rate than pin type, invasive moisture meters. They are less accurate, however, because of their larger sampling area and very limited depth, and because material variations can affect the meter's signals.

Because of their non-invasive nature and quick surveying capabilities, non-invasive meters are the most effective tools for locating abnormal moisture behind and beneath finishing materials such as ceramic tile and vinyl floor coverings. A common misconception associated with non-invasive moisture meters is that they are *non-penetrating.* Although it is true that they do not use a *destructive* penetrating probe, their signal does penetrate the material being monitored. This penetration can be as little as ¼" (6 mm) and as much as 1" or more (25 mm). The depth depends upon the meter make and model, the material characteristics, and in some cases, the meter's setting. Most non-invasive meters penetrate ½" (13mm).

Non-invasive meters will often display measurements on a relative scale. Some models display their readings as a percentage of "wood moisture equivalency," while others simply display it as a "relative" value.

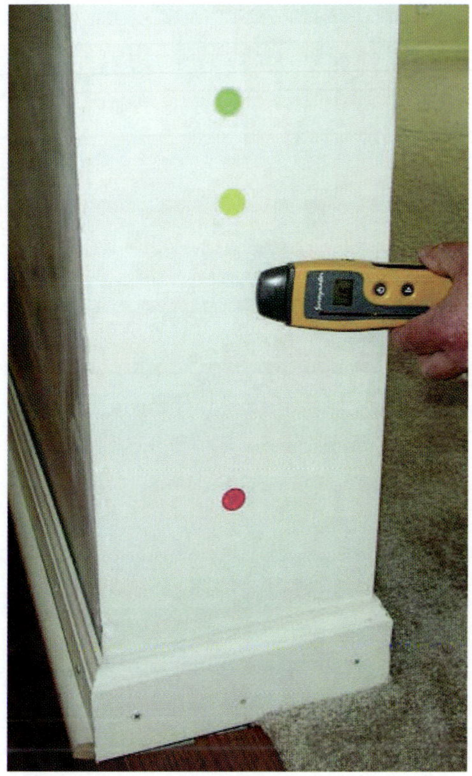

Use the "dot" marking system when using a non-invasive moisture meter.

Particularly when monitoring non-wood materials, it is important to document and report the readings appropriately. Data should be referred to as "relative" or "wood moisture equivalency/WME" if the meter is being used on non-wood materials yet uses a scale that has been calibrated to the percentage of moisture content in wood.

Non-invasive moisture meters can give *false positive readings* on suspect materials for a number of reasons – most often because of metal hidden within or behind the material being monitored. Just as higher levels of moisture change the characteristics of a material, so will metal. Some materials contain chemicals and/or heavy metals that can influence the meter's displayed value. Some common building components that influence a non-invasive meter's accuracy are:

- ☒ Gypsum wallboard (plasterboard) corner bead
- ☒ Metal studs
- ☒ HVAC ducting
- ☒ Plumbing
- ☒ Foil lining (wall paper, insulation)
- ☒ Excessive fasteners (nails)
- ☒ Electrical wiring, conduits and boxes
- ☒ Concrete
- ☒ Heavy metal in paint and paper backings

False negative readings are also possible with non-invasive moisture meters. These false readings result from unseen air gaps behind and beneath target materials. Most non-invasive moisture meters are only capable of carrying a signal through solid materials. The presence of air will limit the signal. Common building surfaces that may present air gaps behind and or beneath them are:

- Wall paneling over gypsum wallboard
- Veneer finish panels on cabinetry
- Swollen baseboard (skirting board)

Applying moderate pressure with one hand while placing the moisture meter on the surface with the other hand will aid in eliminating air gaps behind these materials.

When using a non-invasive moisture meter to monitor gypsum wallboard or vertical surfaces, the meter must be held horizontally. This will allow the instrument to measure moisture in the material at the bottom edge. This is especially important with non-invasive moisture meters that use two rubber pad electrodes. In order for the instrument to detect moisture, both pads must be in contact with the wet material. Holding the instrument vertically will prevent both pads from contacting wet material near the surface edge. Holding the meter horizontally will also allow the instrument's housing to act as a guide to show the actual water level in the wall.

As with any instrument, understanding the limitations of the technology plays a critical role in properly interpreting its output. Reading the owner's manual for each model will provide the information necessary for proper operation and documentation of resulting values. Also, the owner's manual will contain information regarding how often the instrument calibration must be checked.

TIP

Moisture meters have many limitations. To ensure accurate findings, restorers should consult with their meter owner's manual.

Humidity and Temperature Instruments

Thermo-Hygrometer. A **thermo-hygrometer** measures temperature (thermo) and relative humidity (hygro). This information is recorded in the **Record of Drying Conditions** (RDC) and used to calculate other important properties of the air being monitored. These additional properties include:

- Dew point temperature
- Vapor pressure
- Humidity ratio or GPP (g/kg)

TIP

Each of the items above are discussed in detail in *Chapter 7, Drying Science*. It is important to note that the third item, *humidity ratio*, describes the measurement of grains per pound (GPP) or grams per kilogram (g/kg) of dry air. For a better understanding of this measurement, refer to the glossary definitions of *humidity ratio* and *GPP*.

Some thermo-hygrometers are capable of calculating GPP (g/kg) and displaying it digitally on the unit's LCD. Others are capable of obtaining additional information, such as surface temperature, and using this information to calculate other valuable data.
It is important to note that thermo-hygrometers can provide inaccurate information for a number of reasons. Operator error is often the cause while, on other occasions, it is due to instrument limitations. The most common causes of inaccurate readings are:

- Insufficient acclimation time
- Specification limits of the probe being used (Lower cost instruments do not provide consistent readings below 20% relative humidity.)

- Weak battery
- Operator's breath or perspiration
- Proximity of other air sources (HVAC, dehumidifiers, etc.)
- Soiled probe

A thermo-hygrometer is used to monitor any air that can influence the temperature and/or humidity within a restorative drying environment. They are used to verify equipment operation and to evaluate the need for additional equipment, particularly with equipment intended to reduce the humidity and/or control the temperature within a drying environment.

One example of monitoring with a thermo-hygrometer is the measurement of a dehumidifier's humidity removal, or **grain depression**. This type of evaluation is done by taking the temperature and relative humidity of the air entering and exiting the dehumidifier, then using a psychrometric chart to calculate the GPP of each. Technicians can estimate the water vapor removed by the dehumidifier using this difference. This measurement does not provide exact water removal data, because the volume of air flow is not represented. But performance can be tracked and compared as the drying job progresses. Changes in grain depression are used to evaluate the impact of other changes, such as an increase in temperature, or dehumidifier performance. It is important to note the accuracy limitations of the thermo-hygrometer when performing this evaluation.

Thermo-hygrometer readings can be used to assist in additional equipment and strategic decisions made as drying progresses toward the pre-determined dry standard. The table below outlines the number of environments that are measured with a thermo-hygrometer and a simple reason for each:

AREA TO MONITOR	REASON TO MONITOR
Affected Area	Humidity should be decreasing
Unaffected Area	Humidity should be normal and relatively stable
Dehumidifiers	Should be removing moisture
HVAC (if running)	Should be cooling / heating (depending upon intended use)
Outdoor Air	Should be drier than affected (if being used as part of the drying system)

Of the instruments discussed in this chapter, the thermo-hygrometer is one of the most important to calibrate on a regular basis. Relative humidity probes are prone to loss of calibration and should be monitored frequently. Additionally, it is necessary to monitor the meter's accuracy more frequently if the instrument is exposed to abnormally high or low temperature or humidity.

One way thermo-hygrometers are exposed to unacceptable temperatures is when technicians leave them on vehicle dashboards. Freezing temperatures and heat from direct sunlight can adversely affect the meter's set calibration. The operator's manual (or the manufacturer) will provide specific guidelines regarding frequency of calibration for a particular thermo-hygrometer.

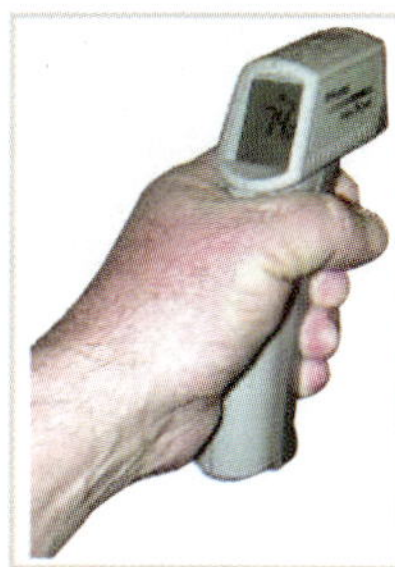

Infrared Thermometer. An **infrared thermometer** is a device capable of reading the surface temperature of most materials without direct contact. Readings are acquired by measuring infrared light, which will vary with the surface temperature of the material.

Inspection Tools and Equipment

Surface temperature readings can be used when evaluating two areas of particular concern to restorative drying contractors:

- Potential areas of condensation and elevated humidity resulting from cool surfaces
- Evaporative cooling

Condensation is not the only resulting problem when cool surfaces are present. Because of the relationship between temperature and humidity, cool surfaces increase the relative humidity in the surrounding air. This impedes the drying process by retarding evaporation.

Measuring differences in surface temperature can actually lead to identifying wet materials. Wet materials will release moisture through the process of evaporation. On a surface where evaporation is occurring, cooling will normally occur as well. Of course, there are limitations with almost any instrument. Major hindrances to the use of an infrared thermometer include reflective surfaces, distance from the instrument to the surface, instrument limitations and vapor barriers.

- *Reflective surfaces* do not emit much IR energy, so they provide very inaccurate measurements using IR thermometers. Also they reflect infrared light emitted from materials other than the target, yielding additional false readings.
- Infrared thermometers measure in a cone shape spreading out from the sensor's tip. Therefore, *the farther the instrument is from the material, the larger the area being sampled.*
- *Vapor barriers* cause false impressions when the user is trying to identify suspect areas. Where vapor barriers exist, little to no evaporation occurs. Even if materials are wet, evaporation may not be sufficient to create cooling measurable with the infrared thermometer.

Infrared Camera. **Infrared cameras** are used for monitoring and inspection purposes. The technology is similar to the use of infrared thermometers described previously, but the camera measures thousands of temperatures across a surface, not just at one point. The temperatures are then translated into a photograph, where colors are used to represent varying temperatures.

Many variations are available as false-color pallettes, ranging from "gray scale" to "rainbow" to show the variation in temperature. Most cameras are capable of detecting differences in temperature as slight as 0.2°F (0.11°C) between pixels, but the absolute accuracy is usually ± 3–5°F.

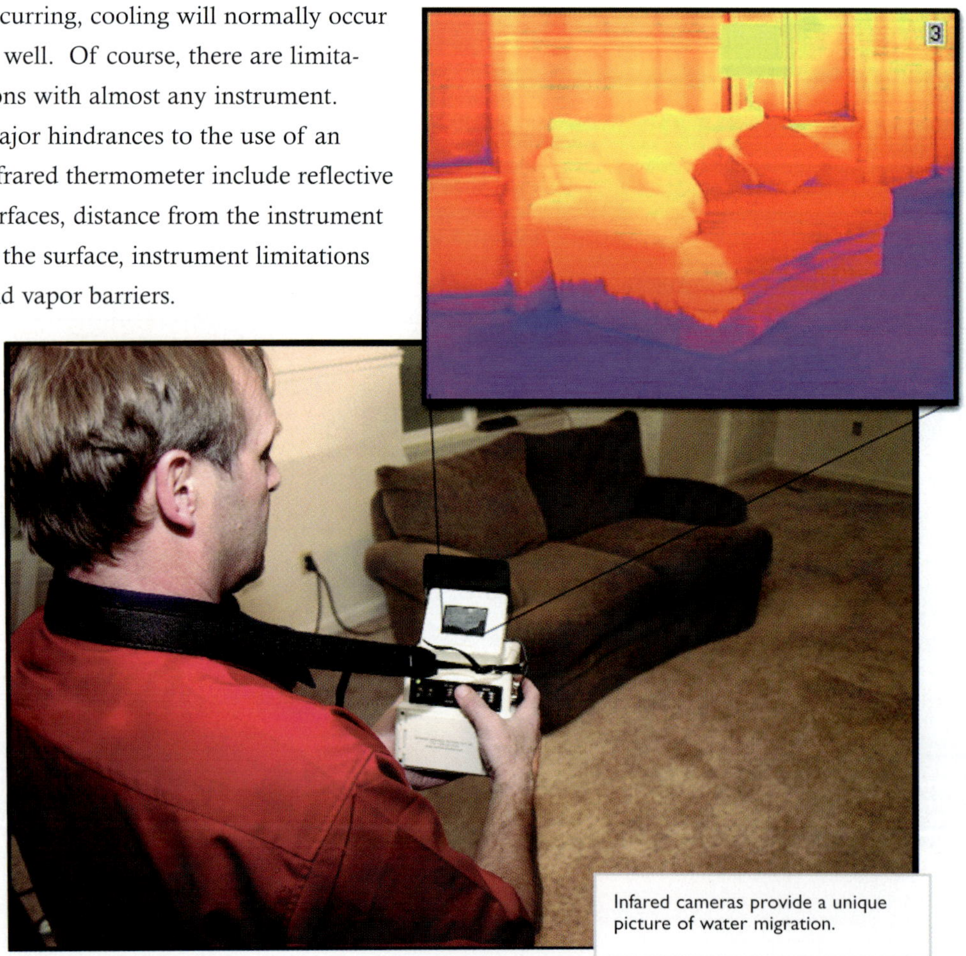

Infared cameras provide a unique picture of water migration.

The primary way in which the presence of water is detected using an infrared camera is through the effects of evaporative cooling. Evaporative cooling, as discussed above, is simply the surface cooling effect of evaporating moisture. Where evaporation exists, cooling typically occurs. This cooling will leave a very evident indication of moisture when viewed through the thermal imaging camera.

Another way in which the presence of water can be detected with an infrared camera is through the effects of **thermal capacitance**. Basically, thermal capacitance is the ability for a material (or substance) to retain thermal energy. Because of differences in thermal capacity, materials (and substances) retain thermal energy for varying amounts of time.

For example, water has a high capacity for thermal energy. When heat is applied to water, its temperature will change relatively slowly.

Therefore, as temperatures change, water can become quite visible. This is most evident as day and night impose very opposite temperatures on building exteriors. This will result in several opportunities to observe temperature variations resulting from differences in thermal capacitance.

The thermal imaging camera shares its limitations with the infrared thermometer. The camera measures the emission of infrared light, which will vary with changing surface temperature. Other factors also influence the infrared light emitted from the surface of materials.

Limitations of thermal imaging range from reflective surfaces, **heat transfer resistance** (R) values, other temperature influences (such as HVAC plenums, etc.), vapor barriers, air filtration, and any other influence on building component temperatures. Materials may also contain excess moisture even if they are not identified using the thermal imaging camera.

Due to these limitations, a thermal imaging camera is only used to help identify "potential" moisture. The visual image is valuable in establishing a map of water migration during the initial inspection. Measurements with moisture meters are necessary to verify actual amounts of moisture in suspect materials identified with the thermal imaging camera.

Visual Inspection Tools

Boroscope. When evaluating cavities within the structure that are obstructed from unassisted visual inspection, **boroscopes** are very beneficial. Through the use of fiber optic technology, boroscopes allow the user to see into wall cavities, beneath cabinetry, under tub enclosures, and into other hidden areas. The visual inspection process allows for a better understanding of the structure's condition.

Boroscopes can allow valuable access to structural cavitites for visual inspection.

Boroscopes usually require a hole or opening of about ½" (13mm) in diameter through which the tip is inserted. The image is transferred to an eyepiece through a long, flexible neck using fiber optics.

Data logging can provide important job documentation.

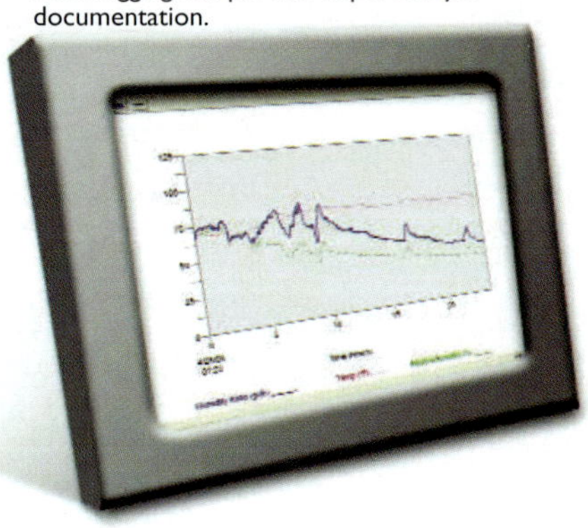

When employing drying systems that will pressurize cavities and therefore create filtration (see **Infiltration, Exfiltration**), inspection of these cavities is important. A visual inspection after a water intrusion also yields valuable information when evaluating water damage, such as shifting of wall cavity insulation.

Many boroscopes are also equipped to work in conjunction with conventional cameras and/or video recorders. Photo records of hidden cavities greatly supplement a job's documentation by showing, visually, how and why the restorer has made strategic decisions related to restoration.

Inspection Mirror. A less expensive but more limited variation of the boroscope, the **inspection mirror** consists of a small mirror attached to the end of a flexible and/or collapsible neck. The mirror allows for inspection of some cavities within the structure, but will require a much more intrusive inspection process if a large opening does not already exist.

Other Monitoring and Inspection Tools

Data Recording (Logging) Devices. Devices are available to log many different types of data, including humidity, temperature, moisture levels and air pressures. In some cases, it is important to capture this data, in both the affected and unaffected areas, between inspection visits.

Data loggers are capable of recording this data at pre-determined time intervals, allowing the restorer to evaluate the conditions present between each visit, or during the entire course of the project.

Data logging devices are extremely helpful in many situations, including when:

- Occupants may cause conditions to change by manipulating drying equipment and/or the building envelope.
- The risk of additional damage is extremely high due to high levels of moisture, an extensive loss area or sensitive high value materials.
- Restoration requires the use of very low humidity and/or abnormal temperatures that may result in additional damage by over drying and/or rapidly drying materials.

In each of these situations, data logging provides essential documentation that is used to evaluate whether or not a condition was ever present that could create potential problems. It also provides exact times that conditions with the potential to cause problems occurred. This information is helpful in identifying the reasons for any problems, such as additional damage or over drying.

In most cases, data logging is not used to make daily drying decisions. Most data logging equipment simply records data that then must be downloaded; it cannot be evaluated while visiting the job site for daily inspection and monitoring visits. Standard instruments need to be used in order to capture current data that can be immediately viewed and deciphered.

Some data logging equipment will report readings in real time through secure internet websites, email, or even through an on-screen display. In these cases, data can be obtained while the device is in operation. When this is true, the data logger serves as a daily decision tool and replaces some handheld instrument use.

Gas Monitors. Another useful monitoring and inspection tool, **gas monitors** are available as either single or multiple gas detectors. Many manufacturers will provide single gas monitors capable of detecting one of 13 (or more) toxic gases, or for detecting oxygen levels in an air sample. Multiple gas detectors monitor levels of two or more gases.

Gas monitors are also available in either standard or maintenance-free models. Standard models need frequent calibration, as often as every six months, while maintenance-free models are disposable. Disposable monitors are typically single gas monitors.

When working in an environment that may have low oxygen levels or high toxic gas levels, a gas monitor is an essential instrument. Specific examples are:

1. When working in a structure that contains combustion appliances.
2. When manipulating indoor air pressures.

High levels of carbon monoxide (CO) will be present if negative air pressures are sufficient enough to draw in air from outdoors through combustion exhaust flutes. This can result in a *risk of suffocation.* Gas monitors assist in identifying this safety hazard – the greatest risk associated with the use of negative air pressures in the restoration environment.

Gas monitors should especially be employed when working around known sources.

Laser Particle Counters. Used to measure the amount of particulate in a sample air mass, **laser particle counters** test the effectiveness of air filtration systems. Laser particle counters are also used to measure the effects of various drying systems on indoor air quality.

Laser particle counters are available in handheld, large portable and remote devices. Price ranges for laser particle counters vary greatly. The factors that have the most influence on the price of the instrument include the particle size detectable, sample time, particle size differentiation, simultaneous measurement of various particle sizes, particle size range(s), data logging features and additional functions (such as relative humidity and temperature).

The most valuable and important specification for use in restorative drying is the ability to measure particles as small as 0.3 microns in size. Many **Air Filtration Devices (AFDs)** contain **High Efficiency Particulate Air (HEPA)** filters, rated to remove 99.97% of particles at this size (0.3 microns). Validating the effectiveness of an AFD can be important, depending upon the nature of the restoration project.

When using particle counters to evaluate AFD effectiveness, readings are taken at the unit inlet and outlet. Differences in particle count are compared to calculate the amount removed. Readings are also taken in the air throughout the drying environment and compared to air outdoors and in unaffected areas of the structure. Major deviation from the unaffected area in particular indicates the need for additional filtration.

Using an anolog manometer

Manometers. Another instrument for measuring the properties of air, the **manometer** detects differences in air pressure between two spaces. Air pressure differentials are particularly useful when working to contain high humidity and other contaminants, and/or to generate airflow in difficult to reach cavities. When using air pressure differentials, accurate measurement is important to determine whether proper pressure differentials have been achieved and are being sustained.

Manometers are available with analog or digital/LCD displays, as handheld inspection instruments, or as data logging devices, and are capable of a wide range of sensitivities. Each of these variations greatly influences the cost of the manometer.

An important specification relative to restorative drying projects is the manometer's sensitivity. It must be capable of measuring differences as little as .02-.04" Hg, or 5 to 7 Pascals. When pressure differentials are used, pressure differences of this magnitude are necessary in order to generate sufficient control over the direction of air exchange and filtration (see Infiltration, Exfiltration).

SUMMARY

The restorer must evaluate the extent of water migration and be able to measure the amount of moisture absorbed by materials in order to properly scope the work to be done. As drying is attempted, the restorer must verify that anticipated progress is actually being achieved. Due to the number of variables present that affect the drying of materials, progress cannot be assumed. Frequent inspection and monitoring is therefore essential.

Proper inspection and monitoring requires a variety of tools and instruments. The most critical of these are are:

- invasive moisture meter
- non-invasive moisture meter
- thermohygrometer

A complete evaluation of the moisture in a structure cannot be completed without these instruments. All other meters and devices in the inspection kit help the technician use these meters more effectively, assist in reaching difficult surfaces or otherwise complement the information obtained by one or all of the three critical meters.

Although the meters themselves must be used, the information they provide can only be understood with proper training. The beginning of this chapter discussed several terms the technician must understand to properly use moisture meters.
The next chapter, (Inspection, Monitoring and Documentation), will expand on these definitions, the use of these critical meters and how to complete necessary documentation.

Chapter 4

Inspection, Monitoring and Documentation

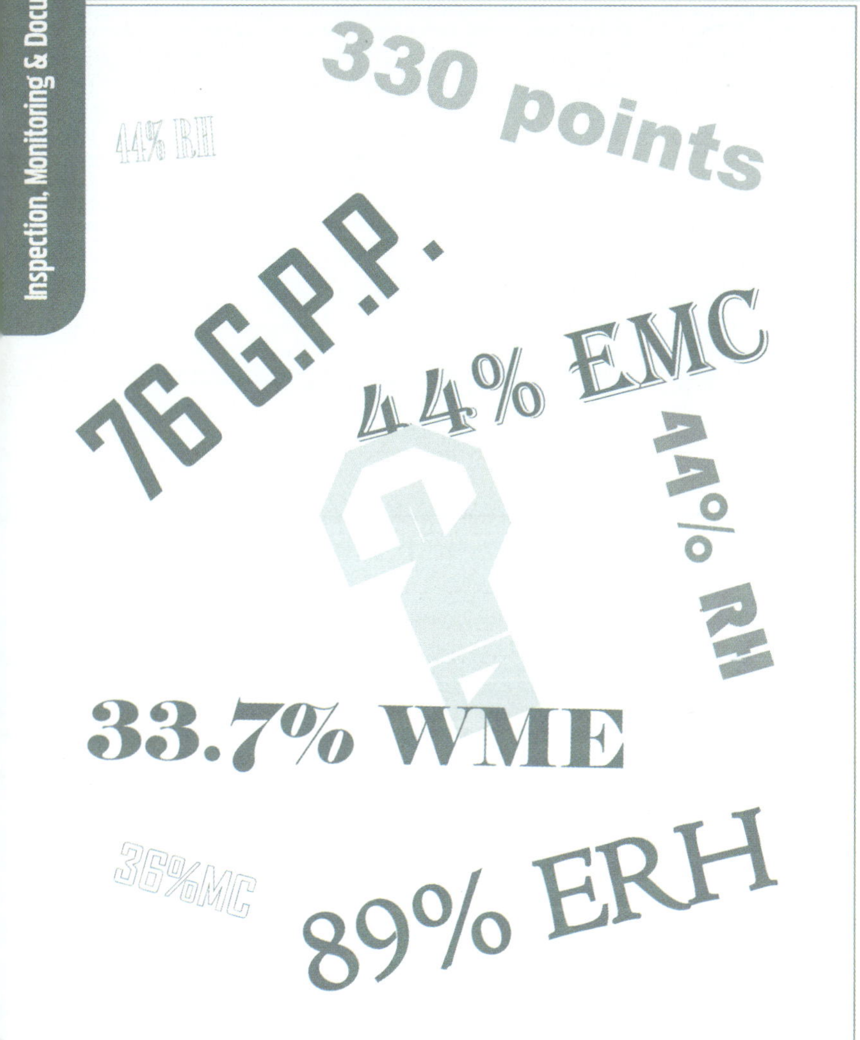

IN THIS SECTION

INTRODUCTION

Restorative drying is a dynamic process in which a number of variables are constantly influencing the restorer's intended result: returning the structure to a clean, dry and safe living environment. The nature of these variables demands that drying environments be closely monitored to ensure that the intended results are realized. This requires a thorough initial inspection to identify items that will influence the restorer's end goal, followed by frequent monitoring to ensure that expected progress is being made.

This chapter discusses the inspection, monitoring and documentation process, divided into five subjects:

1. **Initial Inspection**
2. **Monitoring**
3. **Methods for Monitoring**
4. **Measuring Moisture in Specific Materials**
5. **Reports and Documentation**

Restorers should evaluate structure and contents closely for pre-existing damage.

flooring materials. When this is present, floor (supply) registers should be removed so that all flooring material layers can be noted and duct lines inspected for standing water.

In addition, the type of construction should be evaluated. If the construction assembly poses a complexity which can inhibit the drying process, it should be properly documented. Difficult-to-dry construction can include:

- ☑ Steel studs, which can hold moisture in bottom "C" channel
- ☑ Bay windows, greater number of wall studs, laminated beams, etc.
- ☑ Multiple layers of affected material (retaining more moisture)
- ☑ Vapor barrier or retardant surfaccs/materials

The entire structure is considered in this initial inspection process. This includes substructures, cavities, HVAC systems and any other space or material where moisture may have traveled that could adversely affect the building's appearance, usefulness and cleanliness.

Affected HVAC systems require special consideration. Because of the important function of an HVAC system, additional services may be necessary above and beyond drying. Especially when the internal components of the air-handling unit (e.g., blower, coils) have been directly affected by the loss, a thorough evaluation of the system by a qualified HVAC technician must be performed. Standards and guidelines for cleaning affected ductwork are provided by NADCA (National Air Duct Cleaning Association). NADCA also offers a list of qualified firms.

Identifying Damage

For the purposes of this book, damage refers to a physical change to a material that cannot be reversed by drying alone. It is a permanent condition, requiring physical repair or replacement. There are three types of damage the restorer will need to identify. They are pre-existing damage, primary damage and secondary damage. Each will occur for several reasons, and are discussed separately in this section, beginning with pre-existing damage.

Pre-existing Damage

Pre-existing damage is a condition that was present before the water intrusion that may or may not have been caused by moisture. Because the restoration process can lead to scratching furniture, tearing or delamination of flooring materials, staining surfaces and microbial growth, it is often blamed on the process. When conditions exist that are not a result of the restoration process, yet are not identified before restoration begins, restorers can be held liable for them. It is critical, therefore, that these kinds

Carpet delamination is the separation of primary and secondary backings.

Inspection, Monitoring and Documentation

of pre-existing conditions, or damage, be identified, documented and communicated during the initial inspection.

Restorers need to look for any condition that affects a material or contents' cleanliness, appearance or integrity. Factors to take into account when assessing pre-existing damage include:

- ☑ Quality of building material installation
- ☑ Integrity of building components or contents
- ☑ Appearance of building components or contents
- ☑ Microbial contamination

From a restorer's perspective, the most important forms of pre-existing damage to note are microbial growth (mold) and carpet delamination. Both are commonly associated with water damage and can become the liability of the restoration contractor if not identified and documented prior to restorative drying procedures.

Even if a pre-existing condition was not a result of the latest water intrusion, it may still need to be addressed in the scope of work, especially in cases where the pre-existing condition consists of **microbial contamination.** Information regarding how the scope of work must be altered can be found in Chapter 10. Failure to properly address a pre-existing microbial condition will result in an amplified level of contamination and a spread of the contaminant to other areas of the structure.

If mold is found as a pre-existing condition, it is an indication that water intrusion has happened before and was either not properly remediated or is an ongoing problem. As part of the restoration process, the source of additional water intrusion needs to be identified and corrected.

When **carpet delamination** (the loss of integrity in the adhesives used to laminate the primary and secondary backing) has occurred, further damage is very likely. For this reason, when extensive carpet delamination is discovered, technicians must call it to the attention of all materially interested parties, document the condition and suggest replacement. This communication will prevent the restorer from being blamed for the delamination.

Extensive delamination, although not the result of water intrusion, can be worsened in the extraction and/or drying process. This is because the extraction process applies a physical force to the carpeting and its backings, and depending upon the drying system used (e.g., floating), carpet can be exposed to folding, bending and temperature increase. These factors will promote additional delamination. More information regarding how extraction can cause delamination can be found in Chapter 5, and additional information regarding how drying systems can cause delamination can be found in Chapter 9.

There are a number of causes for carpet delamination – many of them not related to typical restoration activities.

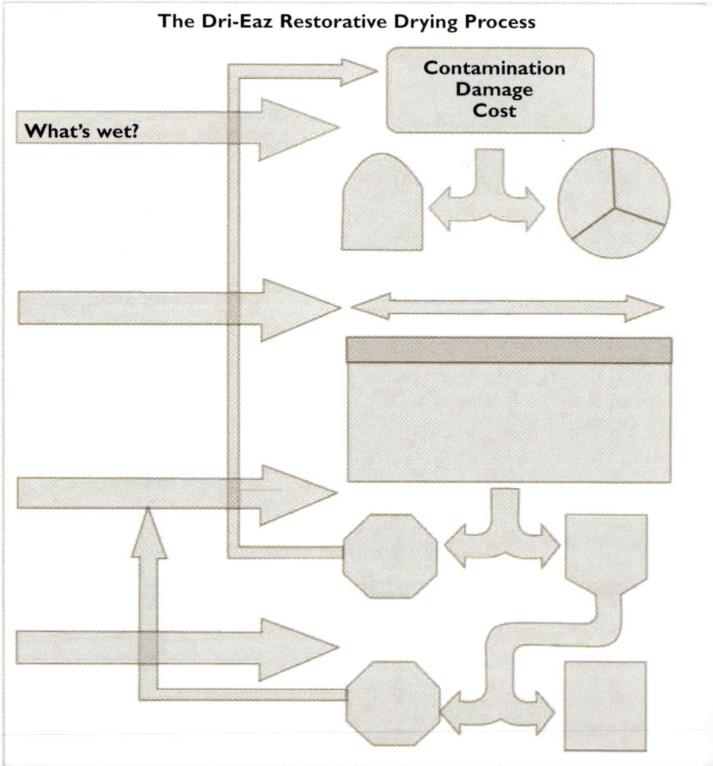

All affected materials must be examined to identify potential damage, contamination and cost of repair.

Typical causes of carpet delamination are:

- Improper installation
- Excessive foot traffic (pivot points)
- Improperly specified carpet pad (underlay)
- Wear from rolling furniture
- Pet damage
- Folding of carpet during removal or drying process
- Improper removal from tack strip (carpet gripper)
- Overly aggressive extraction
- Dragging heavy furniture
- Floating carpet during the drying process

Regardless of the cause of damage, it must be identified and documented in the Initial Inspection Report – then communicated to all materially interested parties. Although this documentation can be written, it is useful to include photographs of the condition.

Primary and Secondary Damage

Water intrusion will affect building materials both immediately and over time. Damage that occurs immediately, by direct contact with water, is referred to as primary damage. Additional damage will occur over time, as water is allowed to remain in contact with materials. Damage that occurs because material remains wet for an extended period of time is referred to as secondary damage.

Water can also affect materials indirectly. Excessive moisture can lead to elevated humidity. Many building materials will absorb excess water vapor and suffer damage, especially when air is allowed to remain above 60% relative humidity. This form of damage from high humidity is also referred to as secondary damage. Common examples of secondary damage from high humidity are:

- Secondary damage when wallpaper begins to release in an affected area
- When doors and drawers swell and do not open properly, and when surface moisture begins to appear on previously unaffected wallboard due to condensation.

Structure and contents must be evaluated for the presence of primary and secondary water damage. As with pre-existing damage, primary and secondary damage need to be properly identified, documented and communicated to the appropriate parties. The scoping process needs to include the necessary work to restore the appearance and performance of all affected material.

In addition, if removal of an affected material is considered necessary in order to properly restore appearance and function, removal of the material will occur in the initial stage of restoration. If a material must be removed, the cost associated with the demolition and replacement cannot be avoided. If the material is removed promptly, the moisture contained in the material will also be removed. This will speed the drying of other affected materials that can be saved, mitigating the total restoration cost.

Hidden damage can occur that will require a more invasive method of inspection. One of the most common forms of hidden damage occurs with wall insulation. When water intrusion results in standing water deeper than the wall plate (sole plate), or when water intrusion occurs from above and results in water running through wall cavities, insulation must be closely evaluated. Inspection access holes are used to ensure that insulation has not compacted and lost its "R" value (Resistance to heat transfer/insulating qualities).

If the wall insulation has been damaged, replacement is necessary. In order to replace most insulation, the wallboard (plasterboard) will need to be removed and replaced along with the insulating material and vapor barrier (if present).

MONITORING

Monitoring vs. Inspection

The monitoring process differs from the inspection process. While an inspection involves gathering data to accurately evaluate the current condition, monitoring involves a series of multiple inspections to accurately evaluate and document changes in conditions. There are many components of the inspection and monitoring process that are critical to the accuracy of the results and the ultimate success of the restoration project.

Inspection, Monitoring and Documentation

CHAPTER 4

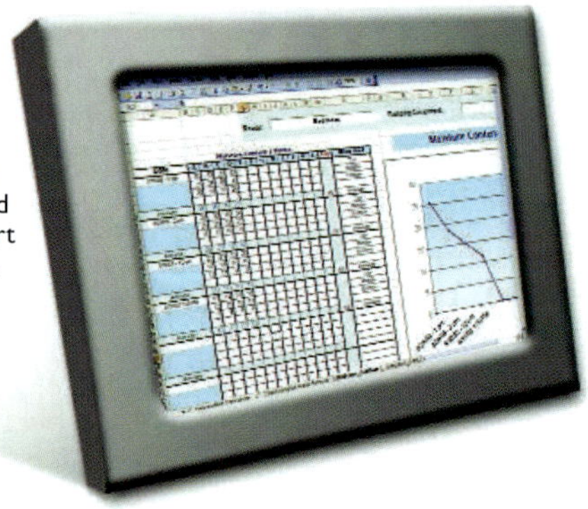

"Restoration Project Manager" can be used to monitor and report daily drying progress.

Inspections must be performed in a way that is easily duplicated, because two inspections can only be compared to evaluate changes if both inspections were performed in the same manner. This requires that procedures for inspection are clearly communicated.

"How Wet Is It?"

Standard Operating Procedures (SOP) are very useful in ensuring that methods used in each inspection are consistent. Knowing how wet a material is requires consistency in the documentation of:

- ☑ Meter used and how it was used
- ☑ Meter setting used
- ☑ Location monitored
- ☑ Depth monitored
- ☑ Method of documentation

It is important to identify the instrument used because meters vary significantly from manufacturer to manufacturer and from model to model. Two inspections of the same material using different meter models will yield two entirely different readings. This leads to inaccurate conclusions and improper drying decisions. Proper inspection and monitoring forms include areas for recording the meter model and settings used.

The Dri-Eaz Restorative Drying Process

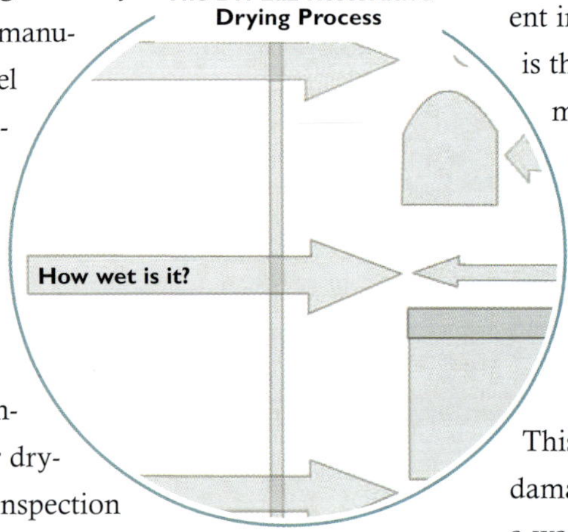

Moisture levels fluctuate throughout a material for numerous reasons, both across its surface and at varying depths. For this reason, it is important to document the exact location and method in which the meter was used. Removable colored dot stickers and painter's masking tape are valuable tools for ensuring consistency in monitoring.

Inaccuracies in the monitoring process also result from differences in and/or incomplete documentation. For example, when recording the height of water migration up a section of wallboard, it is important to note where the measurement begins (from the surface of the carpet, or from the top of the baseboard) – not just how high it is on the wall. All of this information must be documented and/or clearly marked at the inspection point in the structure.

Evaluation: "Is It Drying?"

When inspections are performed properly, technicians can accurately evaluate the progress being made in the structure. Progress is defined as a favorable change in condition.

The Dri-Eaz Restorative Drying Process

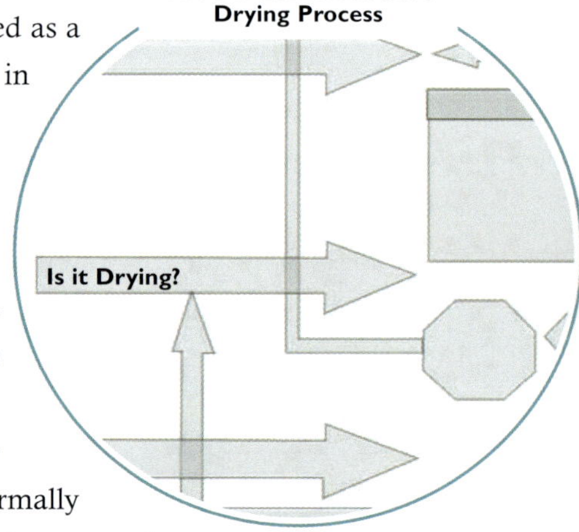

More critical than the amount of moisture present in a structure is the change in moisture over time. Abnormally wet materials are exactly that: abnormally wet. Therefore, the only element that can be changed is the time the material spends in this condition.

This length of time greatly influences the amount of damage that the structural material will suffer due to a water intrusion. Progress is a critical element in restorative drying; slow progress, or no progress, will result in damage.

It is important, therefore, that progress be documented. Monitoring reports differ from inspection reports in that they show the change in moisture levels within a structure. Some documentation systems are designed to serve both purposes in a single report. These reports are discussed later in this chapter.

Monitoring Frequently

Frequent inspection, thorough monitoring and evaluation are necessary to accurately judge progress. This requires daily visits to the job site for most restorative drying environments.

Several factors can hinder progress in returning the structure and contents to a pre-loss condition, increasing the likelihood of complications or limitations. These factors can influence how often monitoring needs to occur:

- ⇨ Size of the loss area
- ⇨ Value and type of materials
- ⇨ Degree of moisture intrusion
- ⇨ Level of contamination
- ⇨ Health of occupants
- ⇨ Structure type and use
- ⇨ Type of drying system/equipment used

Where the risk of secondary damage is greater, monitoring and evaluation need to be performed more frequently. For example, secondary damage is of significance when the restorer is employing large, mobile desiccant dehumidifiers and/or when the loss area is larger than average (200-300 sq.ft/18-28 sq. meters).

On average, most structures need to be evaluated for progress at least once in every 24 hour period. There are several reasons for this practice.

When a structure is evaluated daily, the restorer has the opportunity to perform several tasks that will minimize water damage, equipment usage and overall project time. This results in the return of the structure and contents to pre-loss condition sooner – and a reduction in total job costs.

In a restorative drying environment, conditions are constantly changing. As a result, the type and amount of equipment needed also changes. Just a few examples of what can be done to maximize effectiveness through frequent monitoring are:

- ☑ Repositioning air movement to the wettest areas
- ☑ Removing equipment as materials return to their dry standard
- ☑ Changing equipment technology to meet drying challenges
- ☑ Dealing with limitations and complications

Monitoring more frequently also allows for a greater number of materials to be dried, as opposed to being removed and replaced. If demolition occurs, reconstruction will add substantial cost, time and disruption to the lives of building owners and occupants.

When materials are removed and replaced, it is difficult to meet client expectations. When contractors are able to restore materials and minimize reconstruction and replacement, the majority of job cost is accrued through equipment rental. Greater equipment rental allows for more profitable restoration work.

In summary, frequent evaluation reduces overall job cost, minimizing disruption for the client and/or insurer, and increasing profitability for the restoration contractor. All materially interested parties benefit.

METHODS FOR MONITORING

As the affected structure is inspected and monitored, three methods can be used to ensure accuracy and consistency. These three methods are (1) Average Moisture Contents, (2) Moisture Mapping and (3) The **Dot System**.

Average Moisture Content

Average moisture content is obtained by taking readings in several locations across and throughout a material. The data is then averaged to show the overall moisture level of the material. Because moisture will equalize in many ways and for many reasons, individual readings will increase, decrease and even remain the same while the average moisture level taken as a whole decreases. Average moisture content is particularly useful when evaluating a comprehensive moisture level for all affected materials as well as the drying progress of a particular material.

Inspection, Monitoring and Documentation

CHAPTER 4

Proper inspection and monitoring reports will include the average value for specific inspection points, materials and areas. Examples of these reports are included in this chapter.

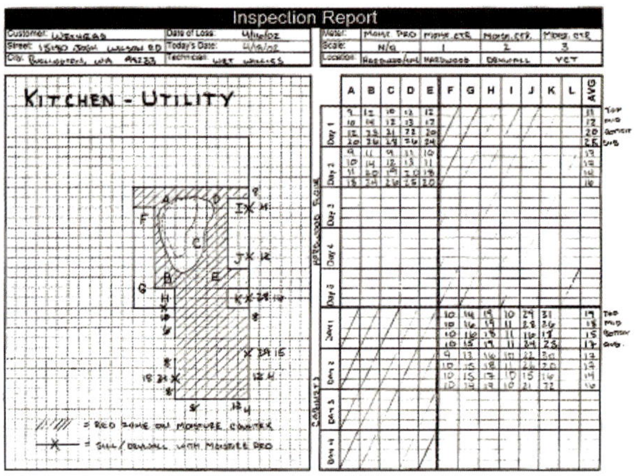

Moisture mapping can provide a guide to affected areas and their progress.

Moisture Mapping

Moisture maps are drawings of affected rooms or areas with corresponding moisture levels for the structural materials in each area or room. These moisture levels are documented in two ways:

- Moisture is documented for an area by defining the location of the inspection point within the area of water migration. Letters are used on the moisture map to record the location.
- Moisture is documented by defining the size of the affected area. Lines are drawn on the moisture map to record the edge of migration.

Lettered locations on the map correlate to letters on a separate or attached document; the attached document records the percentage moisture content or wood moisture equivalency (WME) for each lettered area. Map locations are recorded on the physical surface within the structure and on the paperwork. This is accomplished on hard surfaces and on soft finishing materials (e.g. carpet) using a removable painter's masking tape.

Moisture mapping is useful in identifying the actual size of the affected area and the locations where instruments are being used to monitor progress. Obtaining readings in specific, marked locations ensures consistency by allowing subsequent monitoring to occur in the same location.

The Dot System

The dot system employs the use of removable, colored dot stickers to indicate the relative amount of moisture present in a material. Typically, there are either three or four colors used in the dot system. Each color represents either the level of moisture or the intended restoration procedure.

During the initial inspection, a four-color system can be used to clearly communicate intended restoration procedures. Green, yellow and red dots are used to indicate moisture levels in affected materials that will be dried. Blue dots are used to indicate affected materials that are scheduled for removal. The moisture level that corresponds with colored dots will vary depending upon the instrument being used. By general definition, the dot colors represent the following amounts of moisture:

- Green – Within acceptable tolerance of the dry standard
- Yellow – Abnormally wet but not saturated, at risk for secondary damage
- Red – Near saturation, high risk for secondary damage
- Blue – Material scheduled for removal

When using the dot system, moisture is tracked using the size of the area that is wet (i.e., "What is wet?") as opposed to the actual moisture level (i.e., "How wet is it?").

Example: When monitoring gypsum wallboard, the inches (centimeters) of migration up the wallboard are documented:

- Use a measuring tape to measure from the floor surface to the colored dot placed on the wall, and document this height.
- Further documentation could include photographing the walls with the dots in place.

The dot system can also be recorded using lines on a moisture map to show the extent of migration. Regardless of how the system is evaluated, it must be documented. Removable dots placed on surfaces where airflow is present can become loose. This can lead to misinterpretation if not properly documented.

It will help to date the colored dots being used, or number them in relation to the days of drying. This will assist in the evaluation process as multiple dots are used over the course of drying.

Another method for differentiating data from day to day is to overlap colored dots to show progression in a given material. For example, a section of gypsum wallboard may initially be wet up to 8" (20 cm) in height. Taking for granted that the moisture levels were near saturation, a red dot would be placed on the wall 8" (20 cm) from the floor. When updating the dots on the second day, moisture levels will likely change. New dots will be placed over the previous dots so that the initial dot is still visible. This indicates that two days of monitoring have passed, while showing the change (by color) in the amount of moisture. For example, if the material has returned to an acceptable moisture level, a green dot would be placed partially covering the red dot. The result would be two dots, a green overlapping a red, showing a wet material had returned to an acceptable moisture level within two days.

It is possible for a wall to return to a dry standard near the highest edge of migration, but remain abnormally wet near its base. A new red or yellow dot, whichever is appropriate, would then be placed at the new height.

In addition to these three methods for evaluating moisture in materials, restorers must be able to interpret the value displayed by their meter across a variety of materials. The next section will discuss how to apply the most commonly used meters in the restorative drying industry to typical building materials.

MEASURING MOISTURE IN SPECIFIC MATERIALS

The variety of building materials found in most structures presents a challenge to restorers when trying to quantify absorbed water. The meters commonly used are not designed for use on wood rather than for most of the materials encountered. In order to accurately quantify moisture in materials, restorers must understand the nature of the material and the limitations of the meter being used.

This section will discuss strategies and methods for using common meter types on standard building materials. Further information regarding the meters can be found in *Chapter 3 Inspection Tools and Equipment*. Additional information regarding material characteristics and their response to moisture can be found in *Chapter 6 Materials*.a

Throughout this section, measurements will be discussed using the following terms:

Percentage Moisture Content (%MC) – The amount of moisture in a material expressed as a percentage of the material's oven-dried weight. Examples: A material that is 10%MC is holding:

- 10% of its dry weight in water
- or, a 10 lb. (4.5 kg) sample of wood containing 1 lb. (0.45 kg) of water

Wood Moisture Equivelancy (WME) – A relative reading obtained on a non-wood material using a wood moisture meter. The value does not represent the actual amount of moisture in the material. However, a higher reading indicates more moisture than a lower reading. These readings are only valuable when compared to a dry standard.

Dry Standard – A sample reading taken on an unaffected material. This reading is then used as the unaffected control or drying goal. Dry standards are

Inspection, Monitoring and Documentation

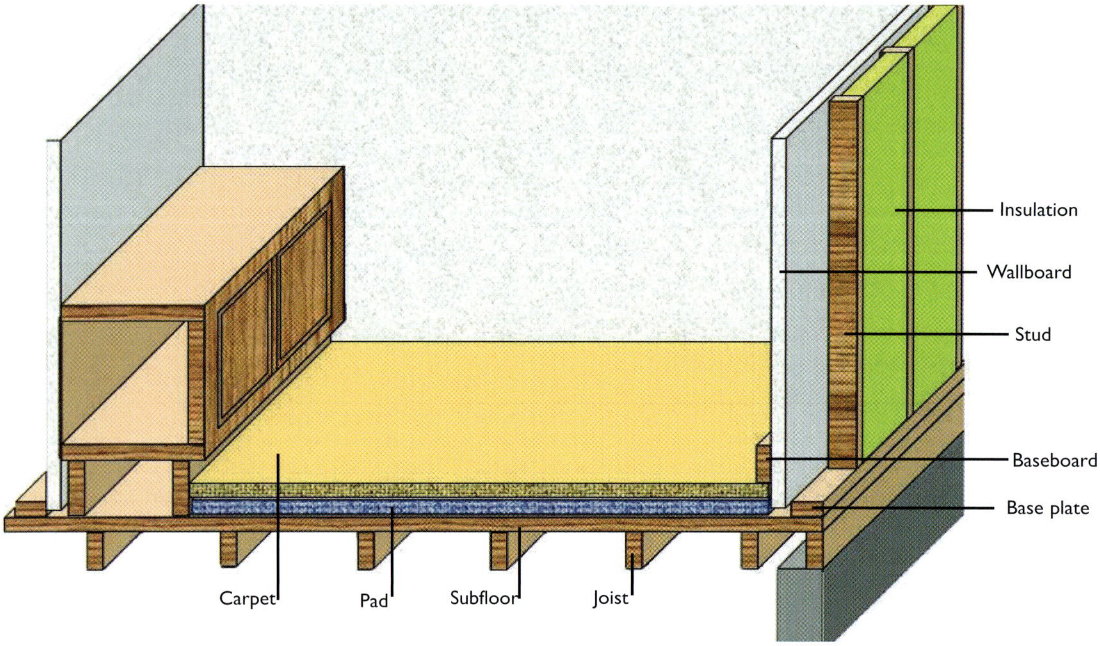

important because most materials will not be monitored by a meter that will quantify moisture in true %MC.

EMC **Equilibrium Moisture Content (EMC)** – EMC refers to the amount of moisture in a material when it has achieved equilibrium with the relative humidity of the surrounding air.

ERH **Equilibrium Relative Humidity (ERH)** – ERH refers to the amount of moisture in air, expressed in relative humidity, when it has equalized with the moisture in surrounding materials.

Structural Wood

Moisture meters used in the restorative drying industry are commonly wood moisture meters. When monitoring wood materials, the value displayed is often an approximation of the actual percentage of moisture content (%MC). As wood is evaluated, there are three significant %MC values that the restorer should be aware of:

16% Moisture contents **greater than 16%** can support microbial growth on many wood materials. Although moisture meters used are not exact in determining actual percentage of moisture content, the number is still valuable as a point of reference.

20% Moisture Contents **greater than 20%** can support the growth of dry rot fungus. If allowed to remain above 20% moisture content, wood can lose structural integrity.

30% The average Fiber Saturation Point for most wood species. The wood fiber has absorbed moisture to the point of saturation. Additional water may be added to the wood material, but it is not absorbed by the fiber itself. It will reside in cavities within cells and other microscopic pockets with wood. Wood that has achieved fiber saturation will appear visibly wet, whereas wood below fiber saturation (less than 30% MC) will appear dry.

Plywood Subflooring. Plywood, especially when beneath carpet and pad (underlay), is evaluated using 2" (5 cm) insulated pins and a hammer probe. Multiple depth readings will yield an accurate value for the moisture in each inspection point. Letters are assigned to each location and recorded on the corresponding moisture map.

Readings obtained at multiple depths are averaged for accurate overall moisture content at each specific location. Specific locations are averaged throughout each room, providing a single overall moisture content for quick and efficient evaluation of area progress.

The edge of water migration is obtained using a moisture sensor and documented on the corresponding moisture map, as noted using lines to show the edge of the affected area.

Plates and Sills. Often overlooked, these structural members can take longer than many other areas within the structure to return to a pre-loss condition. Plates and sills are monitored using 2" (5 cm) insulated pins inserted below baseboards or 6" (15 cm) insulated pins inserted above baseboard.

Insulated pins are an important part of the monitoring process because the carpet, pad, gypsum wallboard or other surface materials contain moisture. If the pins are not insulated, the moisture in these surface materials will influence the meter reading. If neither of these applications adequately reaches the wall plate, 3" (8 cm) wood screws are inserted into the plate or sill. Then, an invasive moisture meter, making contact between the meter pins and the screw heads, is used to obtain the moisture reading in the sill or plate.

Underlayment (Substrate). Underlayment is a substrate material installed beneath a hard, finishing material such as tile or vinyl sheet goods. It is often particle board (chipboard), but can be a number of other engineered wood materials. It is monitored the same as subflooring, discussed earlier. When underlay is beneath a solid finish surface, monitoring processes must be modified to prevent damaging the surface material. Refer to the section titled "Ceramic Tile, Vinyl Composition Tile (VCT) and Vinyl Sheet Goods" for techniques to monitor underlayment and subfloor beneath these materials.

Hardwood Flooring. Hardwood flooring is a common term used for a number of flooring materials. Several forms of wood flooring are installed in homes and other buildings.

Properly inspecting a wood floor begins with identifying the type of flooring material. The types of flooring that are described as "hardwood" flooring include:

- Strip (solid) hardwood
- Strip (solid) softwood
- Strip (engineered) wood
- Laminated wood
- Laminated wood (floating)

Strip (Solid) Hardwood Flooring. When inspecting strip (solid) hardwood flooring, the following key terms are important. They define the properties of each board used to create the flooring material and will influence the way the floor is monitored and how it will respond to moisture.

www.nofma.org

- *Annual Rings* – Most species grown in temperate climates produce visible annual growth rings that show the difference in density and color between wood formed early and formed late in the growing season. The inner part of the growth ring, formed first, is called the "spring wood;" the outer part, formed later in the season, is called "summer wood."
- *Summer Wood* (also referred to as Winter or Late Wood) – Characterized by smaller cavities and thicker walls, which are denser than those in spring wood. Use invasive moisture meters only in the darker area of the wood's grain (created by annual rings). This will simplify the process of repairing the holes.
- *Spring Wood* – Characterized by cells having relatively large cavities and thin cell walls, and lighter in appearance.
- *Quarter Sawn Lumber* – Lumber produced by first

Hardwood floors can sometimes be built atop sleepers above the concrete.

quartering the log and sawing perpendicular to the growth rings. Quarter sawn lumber will generally swell mostly in height, not width. Quarter sawn lumber produces narrow boards, nearly all vertical-grained, and creates more waste, generally making it more expensive than plain sawn.

- *Plain Sawn Lumber* – Lumber cut such that the annual rings appear as flares along the top of the board. Plain sawn lumber will generally swell across its width resulting in a greater risk of damage to the flooring.

In order to effectively monitor strip hardwood flooring, a combination of invasive and non-invasive moisture meters is required.

Non-Invasive Meters – Non-invasive moisture meters allow the restorer to map out the areas of greatest moisture and to identify the extent of water migration. A non-invasive moisture meter will read to a depth of approximately ½"-¾" (1-2 cm) below the surface. This limits the effectiveness of the non-invasive meter's use as a stand-alone meter for wood floors, as it cannot accurately measure the moisture content in both the strip hardwood floor and the wooden subfloor beneath it.

Non-invasive moisture meters are used only to map the extent of migration through the wood flooring. The dot system is helpful for recording the extent of migration, and provides a method by which to document the migration on the corresponding moisture map. Non-invasive moisture meters are also used to monitor the subflooring beneath the hardwood if the subfloor is accessible from below (e.g. basement, crawlspace).

When the non-invasive moisture meter is the only meter used on the wood flooring and subflooring, lettered locations are established and the meter value is

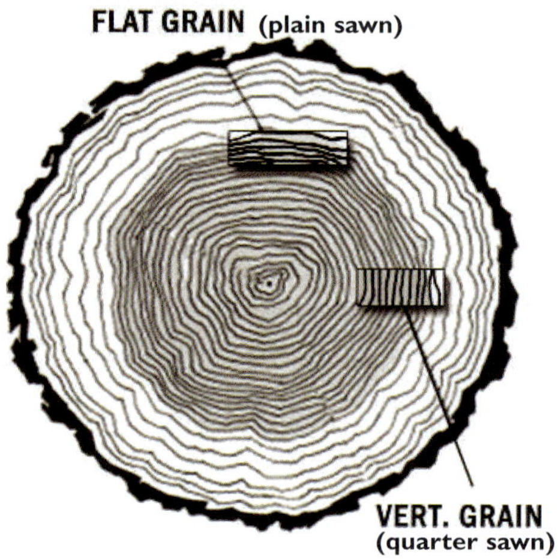

Wood may be quarter sawn or plain sawn.

recorded for specific locations on the corresponding moisture map. This allows for progress to be evaluated during subsequent monitoring visits.

Invasive Meters – Invasive meters enable accurate measurement of the moisture content in the strip hardwood and in the subfloor beneath. Invasive moisture meter readings are mapped using lettered locations recorded in the monitoring report and on the surface of the flooring (e.g., painter's masking tape).

Invasive Moisture Readings – Hammer Probe

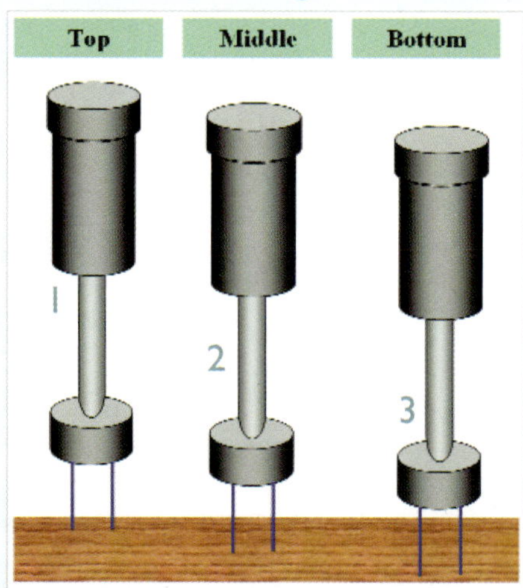

Invasive Moisture Meters can be used to obtain multiple readings in a material, each reading isolated to a particular depth. This allows for very accurate evaluation of moisture levels in the material.

Lettered locations are assigned within the areas of greatest moisture identified by the non-invasive moisture meter. The invasive probes are inserted into the hardwood floor in an area of dark winter wood, near the center of the target board. These locations are more easily repaired. Readings are obtained near the "top" of the wood, the "center" and the "bottom" of the strip flooring, and finally, in the subfloor itself:

1 Obtain the "top" reading by breaking through the surface finish, about 1/16" (2 mm) into the wood itself.
2 Continue, using the same holes, until the tips of the pins are in the center of the strip of hardwood (approximately 3/8" or 10 mm) – document this as the "center" reading.
3 Continue on to a depth of approximately 5/8" (16 mm) for the "bottom" reading.
4 Obtain a final reading by penetrating the subflooring beneath the hardwood (which should be evident by a change in the feel and sound of the hammer probe).

Each depth tells an important story. As this book discusses in *Chapter 6 Materials*, cupping and crowning of hardwood flooring result from moisture content differences between the top and bottom of the board. Cupping will generally be the first indication of abnormal moisture levels as the bottom of the board absorbs moisture. The bottom of the board will then expand, causing a cupped appearance.

Damage is most likely on a hardwood floor when the moisture in the bottom of the board begins to migrate to the top of the wood, causing the board to expand

Repairing holes with a furniture repair kit.

not only on bottom, but also on top. This results in expansion and very high levels of pressure across the floor. This pressure causes crooking, twisting, compression setting and buckling – permanent damage to the flooring that, in most cases, will demand removal and replacement.

Knowing the difference in moisture between the top and bottom of the board provides an educated guess as to whether or not permanent damage has occurred or will occur. When moisture levels in the top of the board are near dry standard, and only the bottom of the board has absorbed moisture, the likelihood of effectively restoring the wood flooring is high. However, when the moisture has migrated to the top of the board, the likelihood of permanent damage is higher.

Hardwood on Sleepers (Screeds) and/or Underlayment above Concrete. Elevated hardwood floors are common in gymnasiums and other athletic environments, in basements and on concrete slab construction. A non-invasive moisture meter is used to map areas of greatest concern as done with strip hardwood floor. The ½"-¾" (1-2 cm) depth range of the non-invasive moisture meter makes it ideal for determining what is wet, and for tracking migration throughout the hardwood, but it will not determine the moisture content of the layer(s) below the hardwood flooring. In some cases, as many as two layers of underlayment are present underneath the hardwood flooring surface. A 2" (5 cm) pin with insulation (a hammer probe) is the ideal tool and will effectively determine the moisture content of each layer of wood flooring.

An additional concern with this type of floor installation is the moisture content of the substrate (usually the concrete floor or slab). If the hardwood flooring is installed over a concrete slab (or suspended concrete floor) and is wet, it is likely that the concrete will also be affected by the water intrusion. Like wood, concrete will absorb and release moisture until it reaches equilibrium with its surroundings. It is important to determine if the concrete slab is affected by excessive moisture. Measuring moisture in concrete is discussed later in this chapter.

Repairing Holes in Hardwood Flooring. There are two important guidelines for monitoring hardwoods; following them both will make repairing floors easy when the job is complete.

1. First, pins are inserted into the dark winter wood to increase the success of repair.
2. Second, pins are inserted into the center of boards and away from edges to avoid splitting damage that will likely occur if pin holes are too close to the edge.

There are two types of repair systems available for repairing monitoring holes in strip hardwood flooring. They include:

- *Burn-In Systems* – Very resilient but with a high level of difficulty.
- *Water Based Systems* – Very simple application, however not as resilient as burn-in systems.

Concrete and Brick

A number of methods are used to evaluate concrete for dry standard. Each method has its advantages and disadvantages, and establishing one method as the standard is difficult due to the varying formulations of concrete found around the world. Fundamentally, there are two methods used for determining the moisture content of concrete that are

approved by flooring manufacturers:
www.astm.org

ASTM F2170 - Relative Humidity Test – This test for concrete provides quantitative results that are valuable when evaluating the potential for mold growth in organic materials placed directly against the concrete slab - a situation common for restorers. Referred to by some as an Equilibrium Relative Humidity Test (ERH Test) or a Water Activity Test (a_W) test, it quantifies the amount of moisture available within concrete.

There are three ways in which this test can be performed, beginning with the most accurate:

1. To conduct the ATSM F2170 test as the standard dictates, a hole is drilled into the concrete slab and, once temperature and humidity are acclimated, the relative humidity within the slab is evaluated. Performing this test exactly to ASTM specification requires several hours to allow the concrete to cool after drilling.

2. A similar test is done by sealing an insulated dome to the surface of the concrete and, once it is acclimated, determining the RH by placing a thermo-hygrometer under the dome. Although not as accurate as the ASTM specified method, this method is more feasible for restoration-related activities. It retains the accuracy of temperature by using an insulated dome. It loses accuracy in relative humidity because the test is limited to the surface of the concrete. Longer acclimation times will yield more accurate results.

3. Another modification of this test is done by sealing a thermo-hygrometer to the surface of the concrete using clear polyethylene sheeting. As with the previous modification, the humidity beneath the plastic is allowed to acclimate with the amount of moisture in the concrete. The thermo-hygrometer is used to measure the acclimated relative humidity. This test is the least accurate of the three because temperature variations between the environment and the concrete will greatly affect the relative humidity reading.

These testing methods are based upon the fact that the moisture in air and materials will continually seek a state of equilibrium. The amount of moisture in a material will be evident by the surrounding relative humidity if equilibrium is allowed to occur. Higher relative humidities indicate greater moisture contents. ERH readings should fall below 60%.

The ERH test method is also used effectively to measure brick, whether performed to ASTM specification or by using an insulated dome. Brick will equalize at a much faster rate than concrete during this test.

www.astm.org

ASTM F1869 - Calcium Chloride Test (CaCl) – This test measures desorption rates. It is conducted similarly to the ERH test above, but the parameters measured are different. A pre-measured amount of hygroscopic salt (CaCl) is placed in a dish, which is placed on a thoroughly cleaned concrete surface. A dome is then placed over the salt and sealed to the concrete. After 72 hours, the salt is weighed to determine the amount of moisture it has collected.

The results of this test tell the inspector the evaporation rate of water from the concrete slab, which is expressed in "pounds" of water, per 24 hours, per 1,000 sq. ft. (93 sq. meters) of exposed concrete. Most flooring manufacturers require that a concrete slab have less than 3 to 5 pounds of water/per 1,000 sq. ft./per day (1.4 to 2.3 kg of water/per 93 sq. meters/per day).

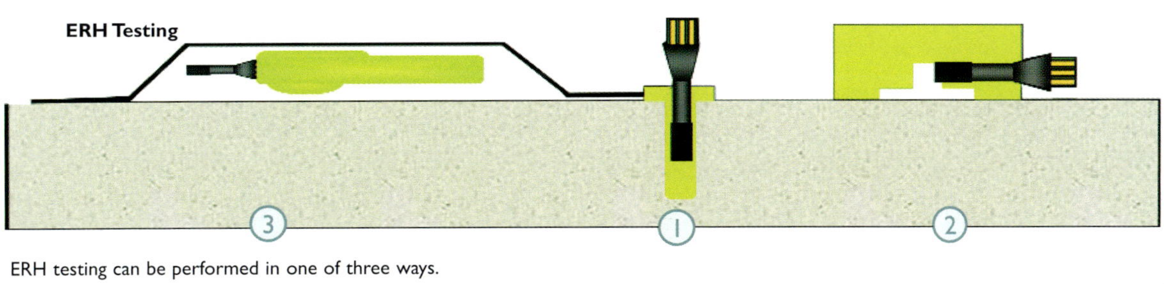

ERH testing can be performed in one of three ways.

Electronic Meters – Wood calibrated non-invasive moisture meters can be used in some cases to inspect, monitor and evaluate moisture levels in concrete and brick. However, because of the multitude of materials used in the manufacture of concrete and the high level of moisture that may be present in concrete even under normal conditions, readings from unaffected concrete may exceed the range of a non-invasive moisture meter. As such, the meter cannot be used effectively when operational range limits are exceeded. When this occurs, measurement technologies such as those discussed earlier (e.g., ERH test, CaCl test) will be required.

There are also non-invasive moisture meters specifically designed for inspecting, monitoring and evaluating moisture content in concrete. However, most models will only measure the moisture present in the top 1/8" (3 mm) of the material.

Pin type (invasive) moisture meters can be used to gain additional information when inspecting concrete and brick. Firm contact between the pins and the surface of the material will measure the relative amount of moisture across only the surface. When high levels of moisture are present, readings will be relatively higher than when the material is at its dry standard.

Plaster. Equilibrium Relative Humidity (ERH) testing is used to measure moisture in plaster construction.

1. Install the relative humidity probe at the surface of the plaster or within wall cavities behind it.
2. Contain the air space using a vapor retardant material such as polyethylene sheeting or a dome to determine the degree of moisture present within the material.

When using this method on plaster, it is necessary to either allow greater acclimation time within the ERH testing containment when surfaces have lower permeability, or to increase permeability of the surface through perforation.

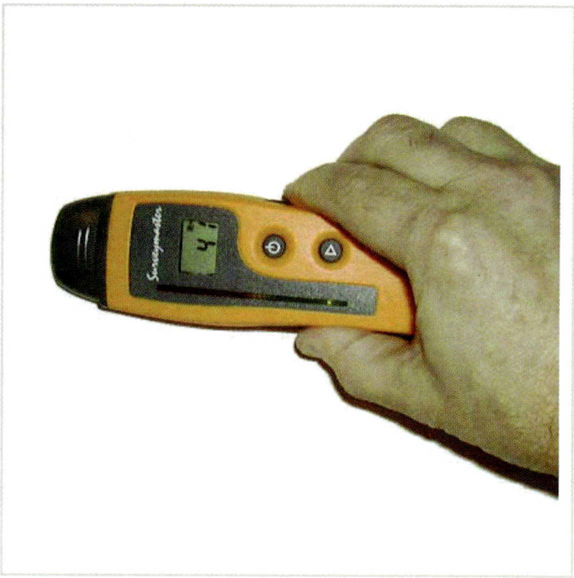

Non-invasive moisture meters are also used to measure moisture in plaster. When using non-invasive moisture meters, readings are relative (qualitative) and only effective when compared to a dry standard.

Invasive moisture meters are also used to evaluate plaster. Short ½" (1 cm) tapered pins are most practical due to the very small pin holes left behind. It is not necessary to insert pins completely if surface readings are high because the pins are not insulated. When surface moisture is detected, it will influence any attempted reading within the depth of the material. This is true because penetration will have to occur through the moisture, and surface moisture will therefore continue to influence readings as pins are forced into the material.

Gypsum Wallboard (Plasterboard)

Technicians primarily use non-invasive moisture meters when evaluating gypsum wallboard. The dot system, in conjunction with a qualitative point value system, is used to accurately measure and track moisture in most gypsum wallboard installations. It is important to measure against a dry standard as most non-invasive moisture meters are only calibrated for wood.

Invasive moisture meters are used to supplement non-invasive readings; this will ensure that moisture is not

Contents should be inspected and monitored, as well as the structure.

trapped beyond the reach of the non-invasive moisture meter. Many installations – such as firewalls, sheer walls, and apartment or condominium common walls – have more than one layer of wall material.

When this is true, non-invasive moisture meters will not accurately measure the entire depth of the wallboard. Verification through invasive moisture metering is necessary in order to ensure that the entire wall installation is inspected.

Contents

The material types present in the item will often dictate the monitoring and inspection methods used for contents. Due to the increased aesthetic value associated with household items, however, most methods will be non-invasive in nature.

Legs, feet, and bases of furniture and other contents that are normally in contact with the flooring are the areas that will take on the greatest amount of moisture. These areas can be monitored using invasive moisture meters. The smallest invasive pin type is used, normally only ½" (1 cm) in length, to effectively evaluate moisture in the very tip of leg posts, feet and bases. Very little pressure is necessary, if any at all, when monitoring and inspecting in this way.

This method is also very effective for monitoring wooden framework behind finish fabric on sofas, chairs, and other upholstered items when holes will not be visible. The relatively small pins will not damage most fabric.

When evaluating contents with hard, visible surfaces, the non-invasive moisture meter is used. Many contents are accessible from multiple surfaces and angles. This allows for a more thorough inspection with the non-invasive moisture meter than when using it for structural inspection.

Although a restorer can only place the non-invasive moisture meter on the outside of gypsum wallboard, it is possible to place the same meter both on top of a wooden table and beneath it. This dual measurement ensures that the entire depth of the wooden surface is evaluated. Regardless of method, it is important that contents are inspected, evaluated, monitored and documented.

Carpet and Underlay (Pad, Cushion)

The first step in the inspection process for carpet and underlay is to quickly assess moisture migration using a moisture sensor. The sensor is the quickest and least expensive tool to track migration through this highly porous material.

Inspection, Monitoring and Documentation

After locating the perimeter of moisture migration, additional monitoring of carpet and underlay is performed only in limited circumstances, especially if technicians can thoroughly inspect and monitor the substrate beneath the carpet and underlay. Because carpet and underlay dry rapidly, progress will be evident in the substrate in relatively short order. Progress in the substrate indicates progress in the materials above it. In addition, a substrate that has returned to its dry standard indicates that the materials above it have also returned to a dry standard.

In cases when the substrate cannot be monitored thoroughly, or when the carpet and underlay may pose a problem in the drying process, ongoing inspection and monitoring of the carpet and underlay are necessary using the moisture sensor or an invasive moisture meter.

When using a moisture sensor, technicians track drying progress in the carpet and underlay (pad, cushion) through a mapping system. The extent of migration during the initial inspection is documented. Then, during subsequent monitoring, the change in the size of the area affected is monitored. Progress is determined by checking thoroughly around the perimeter of the affected area and immediately in the path of airflow from drying equipment. Change in size or surface area affected is noted.

When an invasive moisture meter is used, progress is documented by acquiring a wood moisture equivalency (WME) reading within the carpet backing and/or the underlay. Special care must be taken, so as not to penetrate through the underlay and inadvertently obtain readings from the substrate. To prevent this, pins are inserted until contact with the substrate is evident; the instrument is then pulled up approximately ¼" (6 mm). The point value is documented, along with the exact location of the reading.

Subsequent monitoring visits are performed with the same instrument in the same location.

Ceramic Tile, Vinyl Composition Tile (VCT) and Vinyl Sheet Good

In most circumstances, resilient hard surface flooring can only be monitored using non-invasive moisture meters and a moisture mapping system. Use the dot system to establish the extent of migration beneath the surface of the finish material.

Additional information is obtained by mapping the floor using the moisture mapping and lettering system. Point values from the non-invasive meter associated with the amount of moisture present in specific locations is documented, and then re-evaluated throughout the monitoring process in the same location using the same meter.

This process is particularly valuable when drying occurs slowly. In many cases, the size of the area affected will not change initially. Progress will be evident, however, in the amount of moisture in each specific location. A combination of mapping methods will typically yield the most accurate results.

Wall (Base) Plates

Wall base plates are one of the most commonly overlooked structural components in the inspection process. They are the 2" x 4" boards (5 cm x 10 cm, or similar) installed on the subfloor, on which the wall studs rest. Because of their location, wall plates are commonly affected by migrating moisture, and are relatively difficult to dry. They can only be monitored using invasive moisture meters.

Extended deep penetration probes can be used with many invasive moisture meters, and are often provided with a meter. The length of the probe varies, but is normally no shorter than 6" (15 cm) in length.

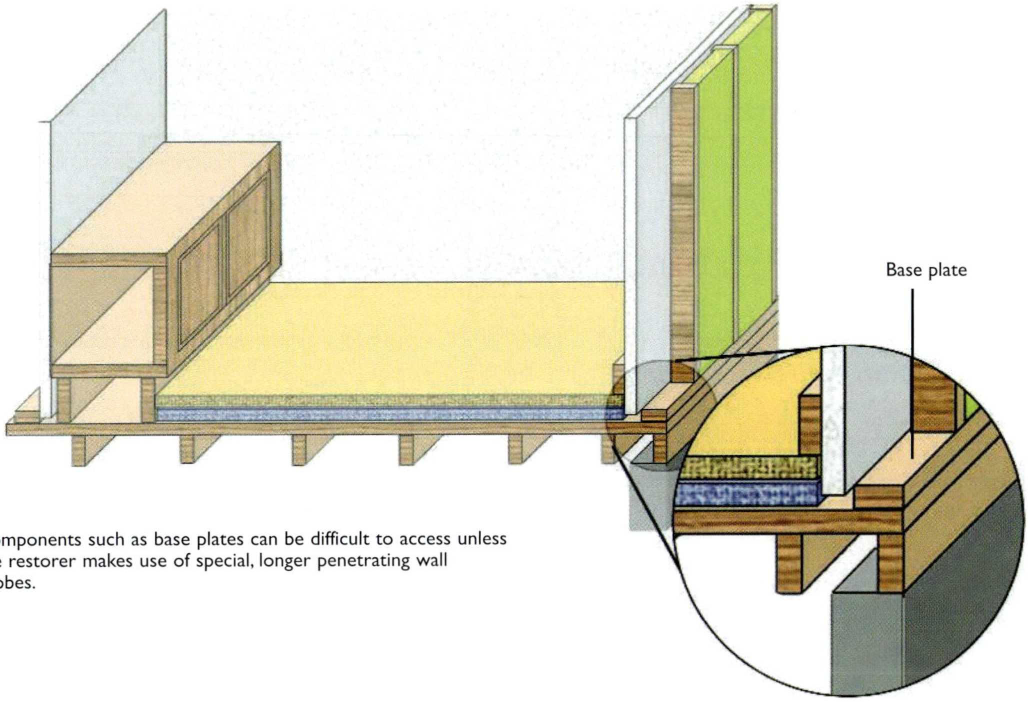

Components such as base plates can be difficult to access unless the restorer makes use of special, longer penetrating wall probes.

These extended probes can be inserted immediately beneath the baseboard, or immediately above, and a reading can be obtained from the surface of the base plate.

More information can be obtained by using two 3" (8 cm, or longer) wood screws which are driven into the core of the base plate. Invasive moisture meter pins can then be placed in contact with the head of each screw. The screw will act as an invasive pin and allow the meter to read deep into the core of the wall plate. Regardless of method, the location inspected and the instrument used must be documented. Subsequent readings must be obtained during each ongoing monitoring visit, performed with the same instrument and using the same method.

Insulation

Another application for the extended deep probe is for inspecting and monitoring wall cavity insulation. Readings obtained are documented and monitored, as are the location and meter type. Whenever affected insulation is discovered, further inspection is necessary to ensure that it has not packed down and lost "R" value (resistance to heat transfer). If it is suspected that the insulation has in fact been packed down, it is necessary to remove and replace the drywall, vapor barrier (if present) and insulation.

Cabinets

Cabinets are monitored in much the same way as contents. The additional steps necessary to monitor cabinets include the use of deep penetration probes to inspect materials behind and beneath cabinets. To do this, inconspicuous areas are pre-drilled, creating holes large enough for extended probes to pass through shelving and other surfaces. In this way, restorers then evaluate gypsum wallboard behind the cabinetry (built in units), substrates (subfloors) beneath the cabinetry and additional structural material that is not otherwise accessible. Because a number of methods are necessary, it is important to document how the cabinetry was initially inspected, and ensure that subsequent inspections are performed in the same way.

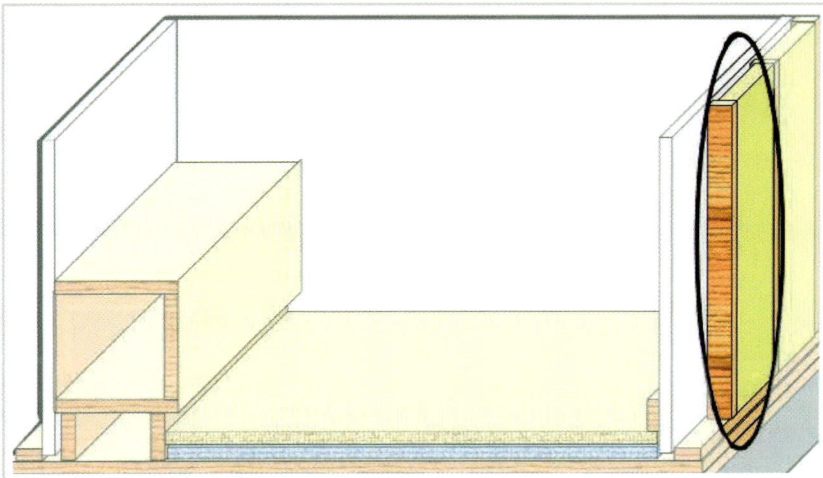

Wall insulation should be checked, especially if standing water was present, or if water ran down from overhead.

Inspection, Monitoring and Documentation

CHAPTER 4

Cabinets present several inspection challenges that require the use of a variety of meters.

REPORTS AND DOCUMENTATION

This section outlines the purpose and content of key inspection and monitoring documents. Examples of these documents are available in the *Appendix*, and on included CD-ROM.

Initial Inspection Report

The initial inspection report, unlike the subsequent inspection and monitoring reports, will capture a variety of job-related details. The initial inspection report is broken into several sections that record general job location information, structure type and use, occupant and owner information, insurance-related information, water source and extent of migration related details, issues related to contamination and damage, and any other influences on the restoration activities.

More specifically, items that are included in this initial inspection report are:

- ☑ All job-related information (e.g. address, client, insurer)
- ☑ Time and date of loss
- ☑ Source of water
- ☑ Types of materials affected
- ☑ Condition of structure
- ☑ Category of water source
- ☑ Category after migration through the structure (if different from category of source)
- ☑ Class of water
- ☑ Pre-existing conditions
- ☑ Details of client conversations
- ☑ Degree of migration
- ☑ Degree of saturation
- ☑ Any other relevant data
- ☑ Name of technician

The initial inspection is intended to document the conditions of the structure and contents at the time of the restorer's initial response. All conditions that might change during the restoration effort are also documented prior to the conclusion of the initial visit.

It is important to use a digital camera (or video) during the initial inspection to record the conditions discovered upon arrival at the job site, including all the information listed above, plus contents placement and condition information. This will help at the end of the job when the contents need to be returned to the property, and will ensure that all contents in the structure have been adequately catalogued for future reference.

Photography also serves as a catch-all, allowing the restorer to capture limitless details that would be impossible to capture with written documentation alone. Photos are also useful when communicating with an adjuster or other party that has not seen the job site, or was not able to see the condition of the structure during the initial inspection.

Inspection Reports

Inspection reports are documents created from an individual visit that capture the current condition of the structure and contents. Inspection reports are like photographs taken at a single point in time. Information commonly captured in an inspection report includes:

- Date and time of inspection
- Instrumentation used
- Job location information
- Current relevant conditions
  - Temperature and humidity
  - Moisture levels
- Name of technician

Inspection reports, although valuable in recording the current condition of materials and contents, do not provide any information on the direction or progress within the drying environment. Because restorative drying by its very nature requires a constant state of change, it is essential to focus on the direction of change rather than on the current condition. For this reason, inspection documents must be formatted to also serve as monitoring reports.

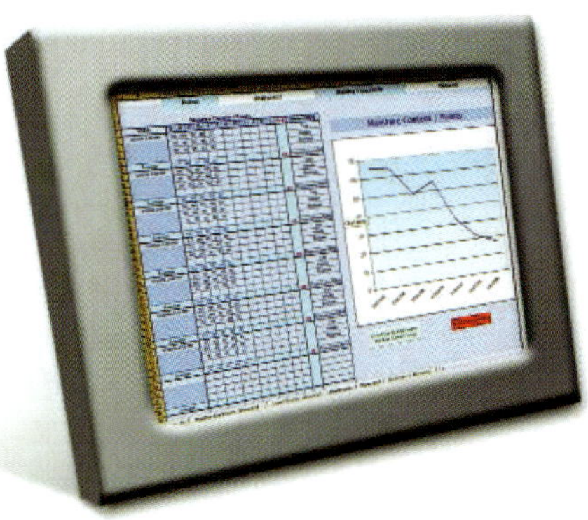

Monitoring reports capture several days of data and clearly show the job's direction.

Monitoring Reports

Monitoring reports are similar to inspection reports, with one very critical difference: monitoring reports record more than an individual "photograph" of the structure and contents and their condition. Monitoring reports are more like time-lapse videos of the conditions in the drying environment, showing the change as drying progresses. Monitoring reports are formatted to include the conditions from the previous inspection, the current inspection, and perhaps, the amount of change between each.

Monitoring reports should include, as a minimum:

- Date and time of monitoring visit
- Job location information
- Instrumentation used
- Previous relevant conditions
  - Temperature and humidity
  - Moisture levels
- Current relevant conditions
  - Temperature and humidity
  - Reduced moisture levels
- Name of technician

Monitoring reports are designed to record the change in moisture conditions throughout the drying process. This change, or lack of change, is then used to evaluate the effectiveness of the drying systems being employed. There are two primary types of monitoring reports:

- *Record of Moisture Levels* – A report designed to track the levels of moisture in affected structural materials and contents over time.
- *Record of Drying Conditions* – A report designed to track the atmospheric conditions in any and all areas that may affect the drying process.

Due to the variety in structures, moisture meters and structural materials, formats for capturing this information in these forms can vary. Examples are discussed in the section "Using Restoration Project Manager" and are provided in the *Appendix*.

Final Inspection Report

The final inspection report serves the purpose of recording the work that has been accomplished through the structural drying effort. At a minimum, the final inspection report captures the condition of previously affected materials and verifies that they have been returned to the pre-determined drying goal, or dry standard.

Additional information captured in the final inspection report includes:

- Date and time of monitoring visits
- Job location information
- Instruments used
- Starting relevant conditions
  - Temperature and humidity
  - Moisture levels
  - Perimeter of affected area
- In-process relevant conditions
  - Temperature and humidity
  - Reducing moisture levels
- End-process relevant conditions
  - Temperature and humidity
  - Moisture levels in materials
- Dry standard, or drying goals
- Acknowledgment of client satisfaction in the condition of
  - Structure
  - Contents

Because the final inspection report serves to document the final condition left by the restoration contractor, it is imperative that the report be completed to include all areas noted as affected in the initial inspection report.

Independent Third Party Evaluations

In circumstances involving severe structural damage, extensive contamination, or other abnormal conditions that can cause potential conflicts, complications or other complexities, a third party evaluation might be necessary and/or advisable. By definition, an independent third party is a company or individual that is independent of the company performing the restoration services.

The job of the independent third party is to:

- Identify initial conditions that may be outside of the restorer's area of expertise;
- Evaluate changes in the structure that may affect the health and/or safety of the occupants during or after the drying process;
- Develop a Scope of Work relevant to services other than the structural drying (mold remediation, asbestos removal, etc.); or
- Act as the project manager for the owner of the property to verify the work being done is in accordance with the Scope of Work produced for the job.

(The *S500 Water Damage Restoration Standard and Reference Guide*)

A third party or independent evaluation is requested by the restorer when it is necessary to ensure that they are following standardized or legislated work practices or procedures, or when performing tasks that require specialized expertise for measurement or evaluation. It is in the best interest of the restorer to consider these services at any point where this is the case.

A third party evaluation may be requested by a party other than the restorer, or may be required by federal, state, provincial or local law. Any such independent evaluation and the reasons for its use should be clearly documented, and the role that the third party will play in the decision-making process must be well defined.

Using the Restoration Project Manager (RPM)
The need for better documentation has driven restorers to record more information when drying affected structures. This has led to many problems. The more data the restorer collects, the more difficult it is to produce understandable, readable documentation. The more time that is necessary to create this documentation, the less likely the restorer is to actually do it. This section explores one option available today that will dramatically simplify documentation, while producing a professional, thorough and understandable final product.

Restoration Project Manager (RPM) has been designed to make complete, thorough monitoring simple. It is a Microsoft® Excel® template (compatible with Excel® version 2000 or higher) designed specifically for recording pertinent job-related information during restorative drying of residential and commercial structures. The Excel® workbook has been pre-formatted and embedded with multiple macros and formulas that transform the large amounts of data into simple graphs and easy-to-follow averages. The spreadsheet is included on the CD-ROM.

Project Manager Cover
The document has been formatted to include several "tabs," each designed to focus on a key area of documentation. The first tab in the spreadsheet is titled "Cover," and has been set up with an area to incorporate a business logo, job name and address and other important company identifying information. This sheet has been set up to auto-fill using information from other parts of the document.

Contact Information
The contact information sheet is set up to capture all relevant contacts for a particular restoration project. The cover tab will use this information to auto-fill the cover page. Here, the insurance company, mortgage company, occupants, property owner, property maintenance company or any other important contacts can be captured.

Project Notes

The project notes sheet is used to capture daily notes relevant to inspections, phone calls, etc. The page has been designed with an "add sheet" function, used to add a page for each day the project progresses. Ultimately, there may end up being 20 or more project notes tabs by the time the job is complete, forming a diary of events and conversations, adjustments to equipment and other pertinent information for future reference.

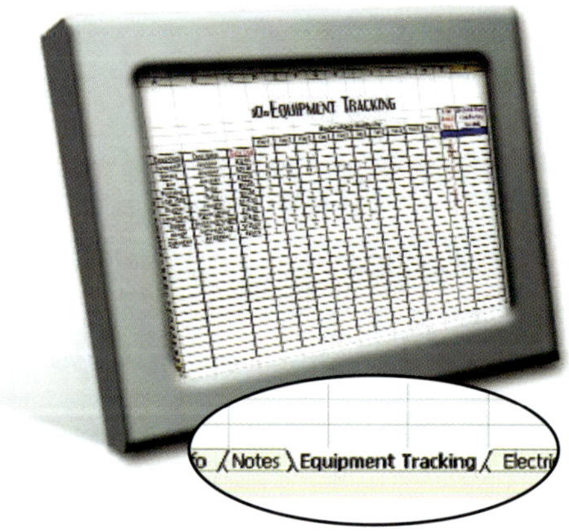

Equipment Tracking

The equipment tracking sheet records the drying equipment used on the loss, by line item and by daily quantity, and allows the user to set the daily rental rate for each equipment line item. The data from this page is used to auto-fill subsequent pages that calculate energy consumption and total job costs. Equipment is recorded by category (air mover) and by model (Sahara Pro X3). Units are then added daily by total pieces installed. The sheet automatically calculates the total rental days for the project, equipment totals by type and a sum total for equipment rental charges.

Outside Services

The outside services sheet records contracted services and their subsequent cost. Markup for overhead and profit can be generated automatically, and can be adjusted to fit the local laws, regulations and pricing. Information from the outside services tab is used to auto-fill subsequent pages of the spreadsheet, such as the total job costs on the Summary tab.

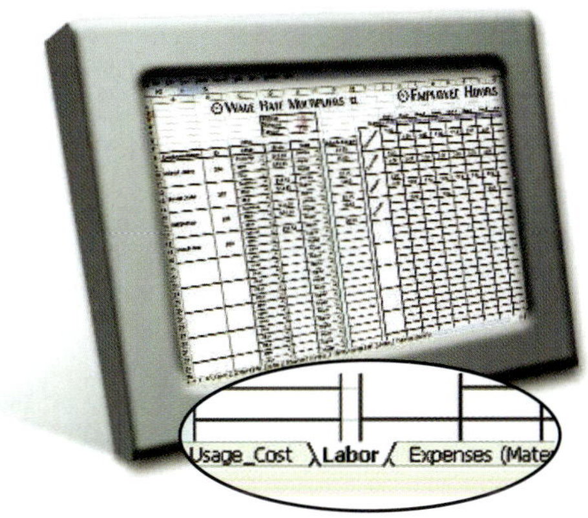

Wage Rate Multipliers & Employee Hours

The "Labor" sheet records the employees used on the project, their wage rate, wage multiplier (emergency, after hours, etc.), their total hours by day, and total wages. This information is also used in subsequent sheets to calculate total job costs, just as the outside services and equipment tracking.

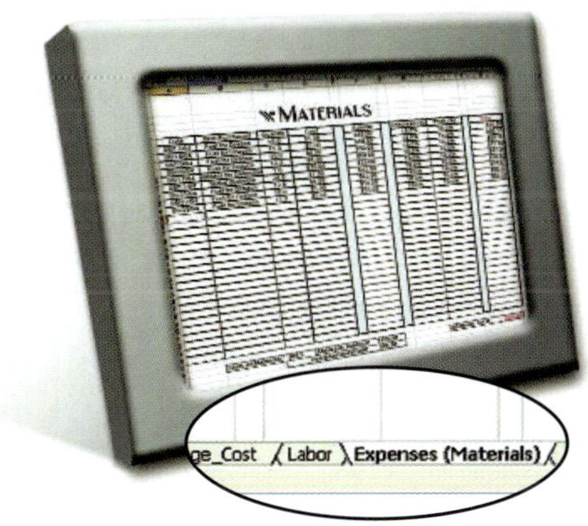

Materials

The "Expenses (Materials)" sheet records the materials used for the project. The sheet includes an adjustment for overhead and profit that can be modified to fit local laws, regulations and pricing. The materials are recorded by line item, unit of measure, quantity, cost and total cost. This information is used to auto-fill subsequent sheets and calculate total job costs.

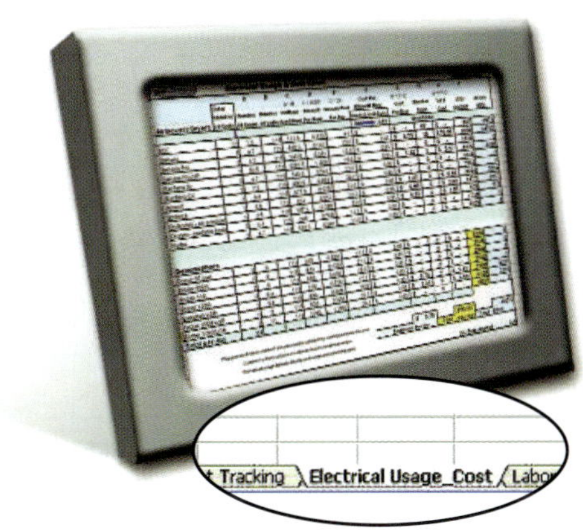

Electrical Usage & Cost Chart

The "Electrical Usage Cost" sheet uses the data from the "Equipment Tracking" sheet to estimate power consumption cost. The sheet can be modified to add amp draw for each line item entered on the "Equipment Tracking" sheet. The sheet also includes a window for entering local power rates and voltage. Energy consumption is then calculated and expressed as a total cost (Kilowatt hours [KW] multiplied by local rate) and British Thermal Units (BTU) per hour (to determine air conditioning needs, when appropriate).

Inspection, Monitoring and Documentation

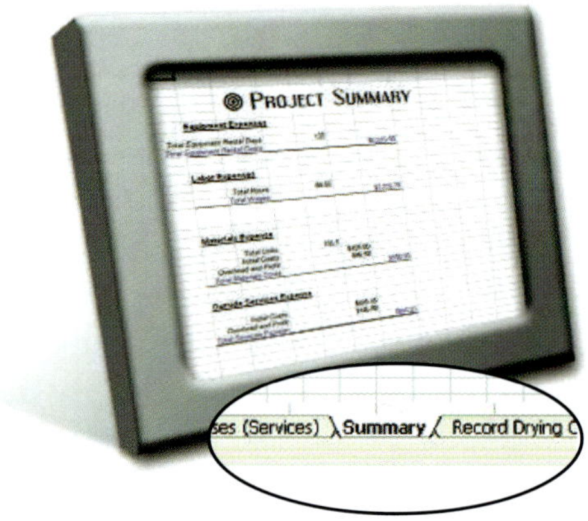

Project Summary

The "Summary" sheet summarizes financial information from the previous sheets (equipment, materials, outside services, labor), offering an overview of current project costs. The sheet includes an area to enter an estimated cost to compare actual costs against. The project summary sheet is the final cost related sheet in the document. All subsequent sheets are designed to track conditions in the restorative drying environment and report progress over time.

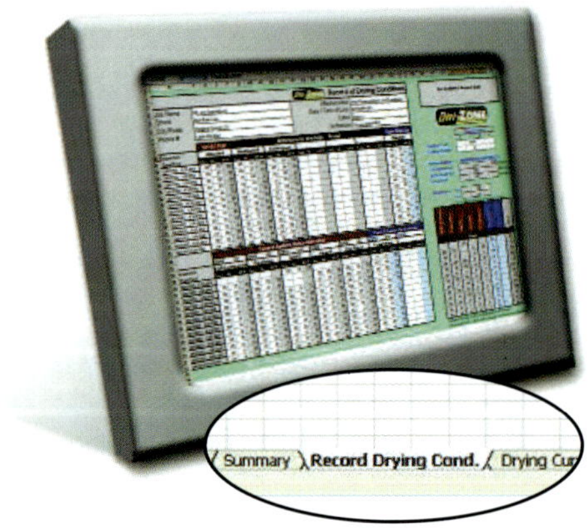

Record of Drying Conditions

The record of drying conditions records the Fahrenheit temperature and relative humidity by date and time across a number of inspection points. The inspection points can include areas of the structure (affected area, unaffected areas), outdoor air, and the output from mechanical systems such as ventilators, air conditioners and dehumidifiers. The sheet uses the data to calculate specific humidity, expressed in grains per pound of dry air (GPP). This is in turn is used to calculate water removal from mechanical dehumidifiers and ventilators, expressed in pints of water per 24 hours. Calculated data is used on subsequent sheets to generate graphs showing progress and change.

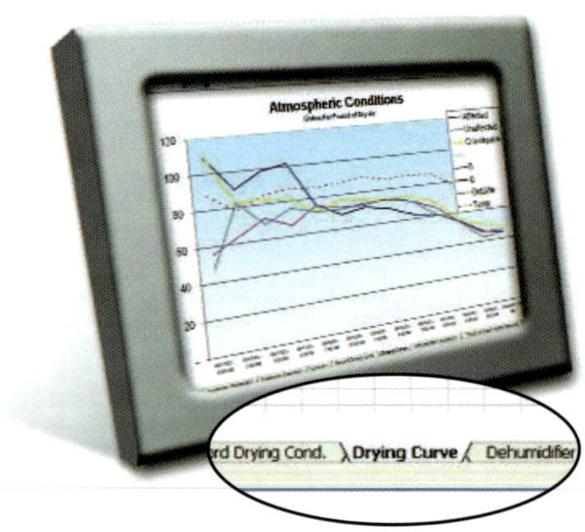

Atmospheric Conditions

The atmospheric conditions sheet ("Drying Curve" tab) is an automatically generated graph illustrating the specific humidity (GPP) in each area monitored. The temperature of the affected area is also plotted on the chart. This is done so that the effect of temperature change can be evaluated. A legend in the top right corner of the graph shows the area represented by each colored line.

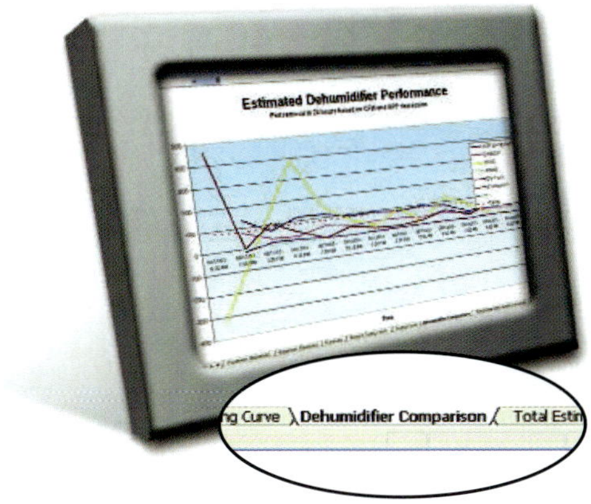

Estimated Dehumidifier Performance

The estimated dehumidifier performance sheet ("dehumidifier Comparison" tab) uses data from the Record of Drying Conditions sheet to generate a graph of water removal. The removal is expressed in pints per 24-hour period. Each piece of equipment monitored is plotted on the chart, as well as the affected area temperature. This allows for evaluation of equipment effectiveness and the impact of temperature on each piece of equipment.

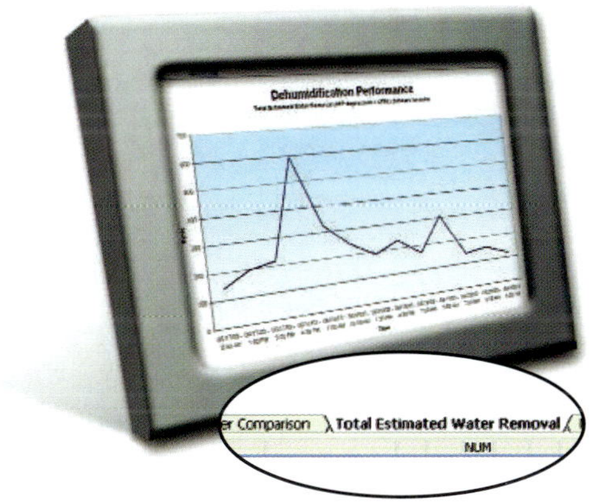

Total Estimated Water Removal

The "Total Estimated Water Removal" sheet is an automatically generated graph. It shows the sum total of water removal by all equipment, expressed in pints per 24 hours. No legend is included with this graph because it tracks a single set of data, referenced by the graph's title. As with each graph in the document, the date, time and scale are all automated and based upon data from previous sheets.

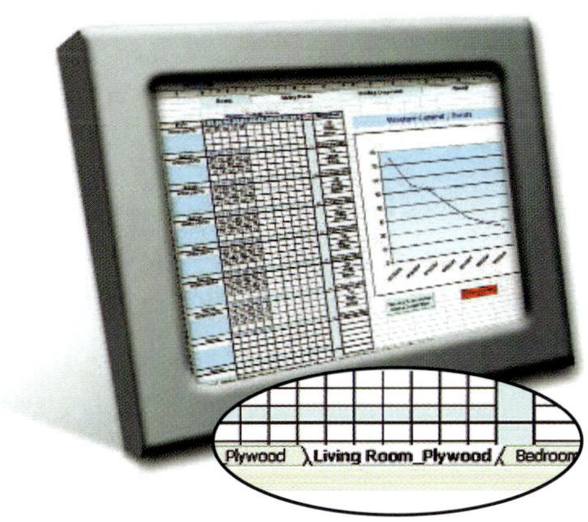

Moisture Content / Points

The room and material sheet records the moisture level in a number of building materials at multiple depths and automatically generates a graph showing change in the average overall moisture content of the target material. The tab name will change depending upon the names entered for "Room and Material" in the provided windows on this sheet. A button on the lower right hand side of the sheet allows the user to add additional sheets, one for each material and/or room. As the sheet is added, the user can then label the room and material, thereby renaming the tab accordingly. The sheet also records the meter used, setting of the meter (if applicable) and the depths monitored (if applicable).

Chapter 5

Extraction

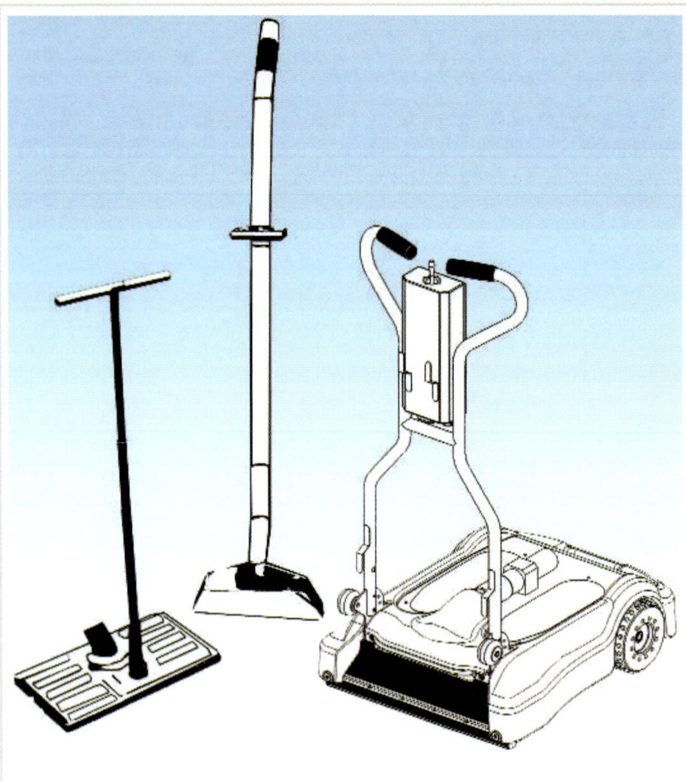

IN THIS SECTION

TIP

Before beginning the drying process, it is critical to:

1. stop the source of water and

2. remove as much water as possible before beginning the evaporation phase of the drying process.

Failure to do either of these will prevent the restorer from mitigating secondary damage.

INTRODUCTION

Clients judge effectiveness of a service by its quality, price and timeliness. Most restorative drying contractors have no difficulty drying structures, but the drying process simply isn't fast enough for many customers.

The most effective way to speed the drying process is to remove as much of the water in a liquid state as possible during the extraction phase. Poor extraction will significantly slow the drying process.

Extraction and evaporation (with dehumidification) are the only ways water can be removed from a wet structure. During the extraction process, liquid water is vacuumed, mopped, squeegeed or otherwise removed from the structure. This takes some energy, but not nearly as much as evaporation.

Water removal by evaporation (with dehumidification) requires that liquid water in the structure or contents first be vaporized – an extremely energy-intensive process. During evaporation, the water is turned into vapor, and then enters a dehumidifier where it collects as condensation and is pumped out of the structure – also an extremely energy-intensive process.

Consider how much energy would be required to remove water from a 50 gallon drum by evaporating it, using an air mover to deliver air and a dehumidifier to capture the evaporated water. How long would they need to run? How much power would be consumed over those several days?

Then, consider the amount of time necessary to extract the water from the same container, using a vacuum powered by a four-cylinder gasoline engine (truck mount). Over the minutes necessary to remove the water, how much gasoline would be consumed?

The difference is quite staggering. The amount of time and energy to remove water physically, in a liquid form, is significantly less than the amount of time and energy required to evaporate it.

While a thorough extraction dramatically increases the speed of drying the structure, it also has other benefits. When combined with proper contents protection, it also helps prevent furniture and rust stains. Thorough extraction also:

- serves to minimize potential damage
- contains migrating water and helps prevent secondary damage
- reduces drying time

Extracted water is usually disposed of into a sanitary sewer. Depending on local authorities and regulations, it might be necessary to dispose of waste water at a licensed waste treatment facility.

The most important factor in extraction is not how much water is removed, but how much water is left behind. It is important to remember that every drop of water left behind will add time to the drying project.

Two-step process for carpet and pad extraction:

STEP 1 The primary goal of the **initial extraction** of the affected area is to contain further migration of the water within the structure. This step is accomplished using a *light wand* attached to either a portable or a truck mounted extraction unit.

STEP 2 The **deep extraction** process is accomplished by using a *sealed or weighted extraction tool* attached to either a portable or truck mounted extraction unit.

If deep extraction is not performed, the carpet cushion must be removed to prevent lengthy drying times. Otherwise, excess water will remain in the carpet underlay (pad, cushion). This water must be removed through dehumidification, which prolongs the drying process. It also increases the amount of equipment required to meet drying timelines and drastically increases the likelihood of secondary damage to the structure and its contents.

FINDING THE WATER

As stated in *Chapter 3 Inspection Tools and Equipment,* a simple moisture sensor will reveal the edge of the moisture migration in carpet and underlay (pad, cushion). Efficient extraction begins with extracting only where water has migrated. Finding the edge of the moisture is important in hard surfaces as well, and non-invasive meters are used to determine where that edge is. Finding this edge in hardwood flooring is especially important, because good extraction from hardwood flooring is critical to drying success.

PROTECTING THE CARPET AND EVALUATING EFFECTIVENESS

Once the water migration edge is located, it is important to protect the carpet from contact with furniture. Simple foam furniture blocks, foil, or plastic tabs accomplish this task. Blocking and padding furniture

Simple foam furniture blocks can protect carpet and contents from secondary damage.

accomplishes a number of important goals, including:

- the prevention of color transfer between furniture and carpet (furniture stains)
- the prevention of rust stains
- the facilitating of effective and efficient drying by elevating furniture (especially low or heavy furniture such as bookcases)

If draperies (curtains) or other window or wall treatments are touching the wet carpet, the appearance of items must be documented (photographs are helpful) and they must be removed from the floor. This is most easily accomplished by using a drapery hanger; pick up the curtain, loop the fabric through the hanger, and hang the hook off the hanger from the curtain rod. More complex or expensive window or wall treatments can be handled by the interior designer who installed the treatment or another appropriate specialist. Any draperies damaged by water are removed immediately and taken to a dry cleaner for restorative efforts.

In some restoration projects it is necessary to remove carpet underlay (cushion, pad). This process will require the restorer to disengage stretched-in carpeting.

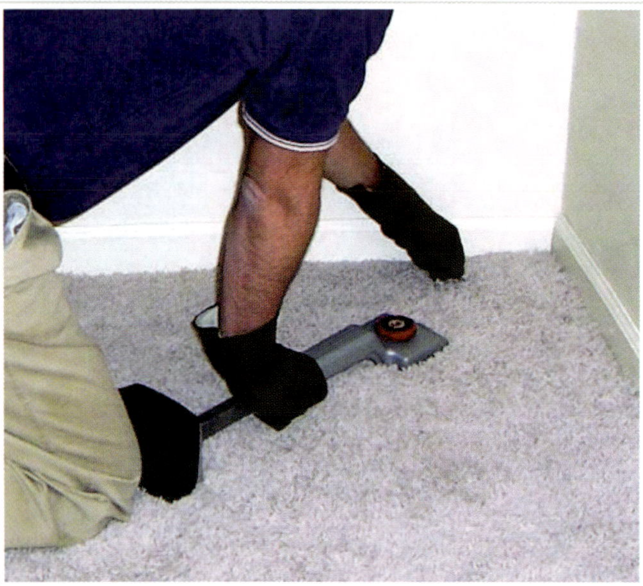

Properly disengaging carpet requires the use of a knee kicker and carpet awl.

Because carpet loses 50%-80% of its strength when it is wet, it is vital to treat it very gently until it dries.

Whenever carpet must be disengaged from the tackless strip, a **knee kicker** and **awl** must be used. The knee kicker is used to release the tension on the carpet and move it away from the tack strip. Then, the carpet awl is pierced through the carpet's primary and secondary backings and used to lift it up and away from the tackless strip. This method of disengagement effectively reduces the possibility of delamination. For personal protection from sharp metal on the tackless strip, technicians need to wear leather gloves while disengaging carpet.

Carpet is disengaged around the perimeter of the room, then rolled back in one direction to expose the underlay (cushion, pad) beneath. It is important to roll the carpet, as opposed to folding it. Folding carpeting adds stress to the primary and secondary backings and can lead to delamination.

If the carpet is to be re-installed, technicians must avoid cutting carpet unless it is absolutely necessary. When carpet must be cut, it is cut along existing seams. This is done while the carpet is disengaged by peaking the two sides of the carpet along the seam, then cutting along the existing carpet seam using a hook blade knife (i.e., vinyl flooring knife). It is also helpful to use a pencil to separate carpet fibers on either side of the seam, allowing for a clean cut along only the seaming tape. This process makes reinstallation and re-seaming of the carpet much easier for the installer. It also increases the likelihood that the client will be happy with the installation after the job has been completed. Cuts must be as straight as possible and must not sever or remove any carpet fibers.

If the carpet will be dried in place, without removing the underlay (cushion, pad), extraction effectiveness is tested by using a knee kicker and carpet awl to disengage the carpet from the tack strip in a corner of the room, and then inspecting the underlay (cushion, pad). The technician disengages and pulls back a 4 sq. ft. (.4 sq. meter) area in an affected corner. The pad is lifted after removing any staples, and squeezed firmly (wearing gloves).

This test is performed after the deep extraction tool has been used on the surface of the carpet. A single pass with the tool is performed, then the cushion is tested. Extraction is considered to be complete when no water drips out of the pad when it is squeezed in this manner. If water does physically drip from the carpet cushion, additional passes are necessary. The carpet is re-engaged, an additional pass is performed, then the cushion is re-tested. It is important to re-test by squeezing the pad in a new area, moving approximately one foot (30 cm) from the area that has already been tested.

EXTRACTION TOOLS

All extraction tools rely upon the principle of airflow through the affected carpet and underlay (pad, cushion) in order to remove water. High negative pressures are created by the vacuum equipment, whether portable or truck mounted. Hoses, usually 2" (5 cm) in diameter, are then used to connect the vacuum equipment to an extraction tool. Airflow through the tool and hose – ultimately returning to the vacuum equipment – removes water from the material and carries it to the vacuum's waste tank.

Restrictions in airflow through the tool, the material, or the hose will impede the extraction process. It is important to do everything possible to ensure that this airflow is not restricted. This is discussed further under "Vacuum Equipment".

Light Wand

The most common extraction tool is the light wand (standard carpet cleaning wand). Light wands come in different lengths, diameters, shapes and sizes, and can be made of different materials, but they all perform the same task: containing the migration of water through carpet and underlay (pad, cushion).

Extracting with a light wand

After the light wand is used to extract water from the surface of a carpet, the weight of water is greatly reduced – slowing down the migration of the water through the carpet and underlay (pad, cushion) and into the structure and materials.

The use of the light wand is the initial step in the extraction process, and must be followed by other extraction tools or other means of physical water removal if underlay (pad, cushion) is to be saved. If no underlay is present, as with direct glue-down carpet, a light wand alone is effective in removing most of the water by passing the light wand several times over each affected area. In cases where underlay is being removed, light wands alone will provide sufficient water removal. Technicians must first extract the surface of the carpet with the light wand, then disengage carpeting and remove underlay (pad, cushion). After underlay is removed, the water is extracted from the carpet a second time.

Squeegee and Squeegee Wand

Hard surface floor squeegee wand

The squeegee wand is the best tool for removing water from hard surfaces. It conforms to a hard surface without damaging the surface or tool head. It is important to note that squeegee wands are not used for standing water. When standing water is present, submersible pumps are used.

In addition to squeegee wands, standard squeegees are also valuable in removing excess water from hard surfaces. They are used to move small amounts of standing water to submersible pumps, floor drains or vacuum equipment.

Deep Extraction Tools

Currently two types of deep extraction tools exist: vacuum sealed and weighted compression. Vacuum-sealed deep extractors make a tight seal with the carpet in order to remove water from the sub-surface underlay and carpet. Vacuum-sealed tools are either mechanical or manual.

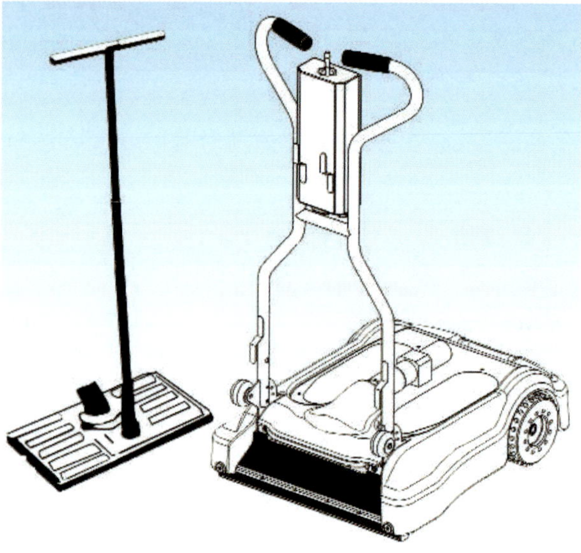

Two examples of deep extraction tools.

Deep extraction tools that are mechanical govern their own speed. When extracting water from large areas of carpet and underlay, consistent speed is critical. Using a mechanical extraction tool increases the effectiveness of the extraction process greatly by ensuring even, consistent extraction.

Weighted compression tools use a combination of weight and a tight seal to remove water from the subsurface, underlay and carpet. Weighted compression tools are also either mechanical or manual.
Carpet construction impacts the effectiveness of extraction tools. Because all deep extractors require a good seal, Berber or multi-level loop carpets are more difficult to extract. Air enters the extraction tool by passing over dips created by the multiple loops or other areas that are uneven in height. Carpet that is thicker or denser with fibers will also be more difficult to extract, as it will restrict airflow.
Similarly, thicker or heavier underlay is more difficult to extract because it also restricts airflow.

Lighter face weight carpet and cut pile carpets are easier to extract. As with any deep extraction, the technician must first test the extraction effectiveness by disengaging the corner and firmly squeezing the underlay (pad, cushion). It is important that the technician continue to extract until no water drips from the underlay.

Vacuum Equipment
The vacuum equipment does the real heavy-duty work of the extraction process. While the extraction tool is actually in contact with the wet surface, the vacuum is what removes water from the environment. Types of vacuums used in restoration range from truck mounted machines powered by gasoline (petrol) engines to portable, electric vacuums.

It is important to note that many small vacuums, such as shop (domestic) vacuums, do not provide sufficient vacuum power to support water removal from carpet and underlay. Restorers must use equipment specifically designed for removing water from these materials.

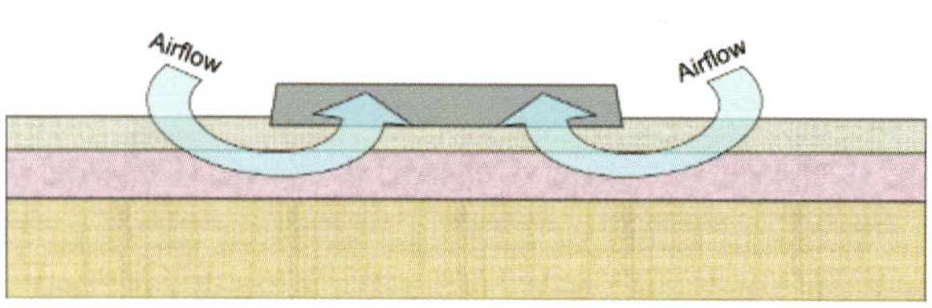

Vacuum-sealed deep extraction tools operate by means of a tight seal on the surface of the carpet.

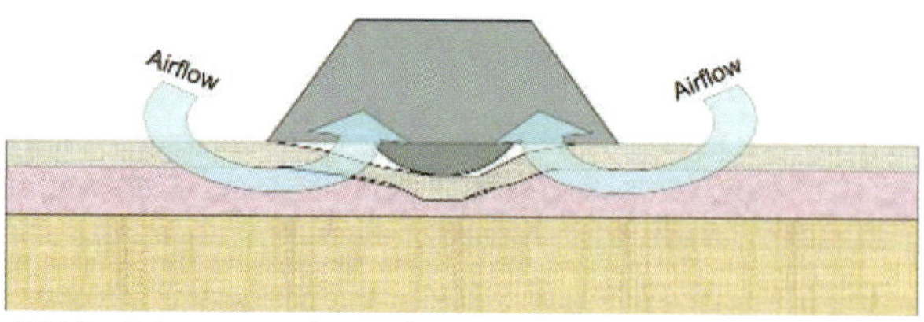

Weighted deep extraction tools use compression to access water in the carpet cushion (underlay).

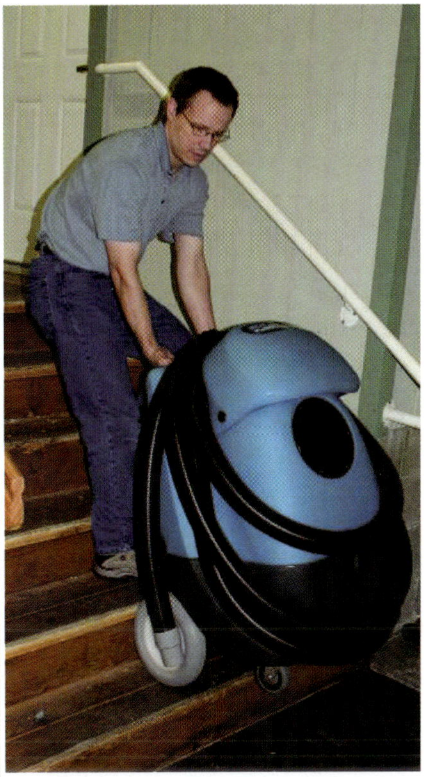

Portable extractors can be more effective than many truck-mounted extractors, especially when hose lengths are shorter.

Regardless of the type of vacuum equipment being used, several guidelines must be followed to increase extraction effectiveness:

- If excessive contents are present in a room, minimize handling and speed the process by extracting the center of the room first. Then move contents to the center of the room, blocking and protecting contents, and extract the outside perimeter.
- The extraction hose must be the shortest length possible, as resistance increases with hose length.
- The extraction hose must be the largest diameter approved for the tool and vacuum equipment (refer to manufacturer's literature). On deep extraction tools, this diameter is 2" (5 cm).
- Loops and bends must be avoided.
- "Live" or "active" hose reels, while great for carpet cleaning, are not to be used for extraction.

In terms of vacuum power, truck mounted equipment generally provides better performance than portable equipment. The limitation associated with truck mounted equipment is the fact that it is mounted to a vehicle and consumes fuel. It must remain outdoors during the extraction process. When operating in a large structure, extreme hose lengths will be necessary in order to reach the affected area. Longer hoses diminish the truck mounted vacuum's performance. In these cases, portable vacuum equipment will provide better results.

A limitation associated with portable vacuum equipment is the fact that it exhausts air from the vacuum blower directly into the air surrounding the machine. When extracting contaminated water, this exhaust spreads contaminants in the indoor environment. To prevent spead of contaminants consider the cleanliness of the water being extracted and the location of the portable vacuum.

Submersible pumps

One of the most valuable and time saving tools in restoration is the submersible pump. Submersible pumps are available at most hardware stores and are relatively inexpensive. They are available in many different sizes and pumping capacities, so it is important to read the specifications for the pump being purchased.

- Smaller pumps remove up to 600 gallons (2,271 liters) per hour, down to ¼" (6 mm) in depth.
- Larger pumps will remove up to 2,400 gallons (9.085 liters) per hour, down to a depth of ¾" (2 cm).
- Gas-powered submersibles are also available that will remove 3,500 to 4,000 gallons (13,250 to 15,140 liters) of water per hour. (However, the disposal hose is comparable in size to a fire hose – meaning disposal of water at the resulting flow rate can become an issue.)

Some water intrusions result in substantial standing water (7" to 8', or about 2 m), but even 1" (2.5 cm) of standing water is significant.

It is important to remember that every cubic foot of standing water present in a structure contains 7.48 gallons (3.8 liters) – and each gallon (28 liters) of water weighs 8.34 pounds (3.8 kg). Combined, these numbers can result in some amazing facts. Imagine a basement that is 20' x 40' (6 m x 12 m) in surface area, and consider the amount of water that would be present:

Standing water

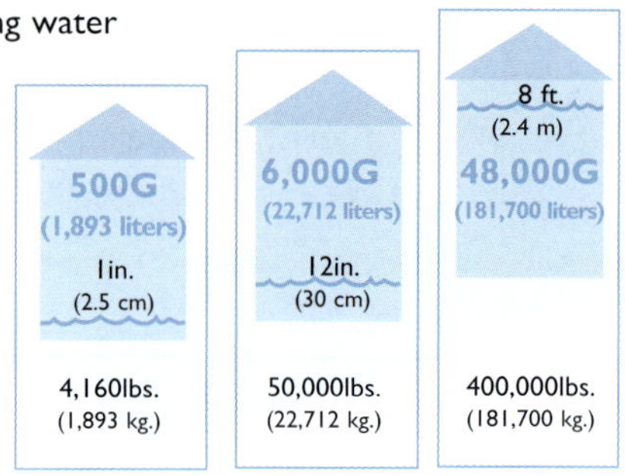

Obviously, a truck mounted or portable extractor would be overwhelmed by this amount of water. In these cases, a submersible pump is a better tool.

To determine the amount of standing water present on a water loss:

STEP 1 Establish the length and width of the affected area (estimate by measuring the outside of the building) in feet.

STEP 2 Measure the depth of the water in feet.

STEP 3 Multiply the length x width x depth to find cubic feet (cubic meters) of water.

NOTE: When measurement includes inches:

* Convert inches into a decimal by dividing the inches by 12.
* For example: 3" divided by 12 yields 0.25’. (Or, 10 cm divided by 100 yields 0.1 meters.)

STEP 4 Multiply cubic feet by 7.48 to find how many gallons of water are present. (Multiply cubic meters by 1000 for the number of liters.)

STEP 5 Multiply gallons by 8.34 to find the weight of the water present. (One liter weighs one kilogram.)

EXTRACTING STANDING WATER

Standing water on the inside of a basement or crawlspace exerts pressure on the walls and/or foundation of the structure. If the water table of the surrounding ground has risen along with the interior water, equal pressure is pushing toward the inside of the structure. This pressure is called hydrostatic pressure.

Removing all of the interior water too quickly, while the pressure still exists outside, can result in structural failure of, or damage to, the walls. To prevent this problem, it is important to calculate the amount of water present and divide that by the flow rate of the submersible pump being used. The result will be the amount of time necessary to remove the water from the area. This information is then used to evaluate the progress while submersible pumps are being employed.

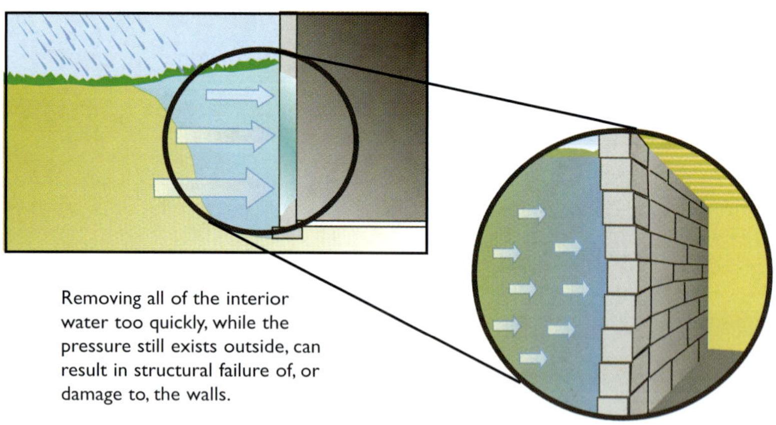

Removing all of the interior water too quickly, while the pressure still exists outside, can result in structural failure of, or damage to, the walls.

TIP

When removing contaminated water:

1. Consider using sewage removal services (e.g., septic trucks) to remove large volumes of contaminated water
2. Ensure portable vacuum equipment is vented outside to prevent spread of contaminants
3. Check with local authorities for proper waste water disposal sites

During the extraction process the depth of water must be monitored to ensure that it is actually falling at the expected rate. If the water is not receding at the expected rate, it is possible that a high water table exists outside the structure, and water is entering due to hydrostatic pressure. Pumping efforts must be halted entirely, until the water table recedes outside and/or a structural engineer or other appropriate expert has been consulted.

EXTRACTING WATER FROM HARDWOOD

When hardwood is being dried, it is necessary to extract it thoroughly, just like any other surface. This is accomplished using hardwood floor drying mats and a portable vacuum. Mats are placed on the floor and attached directly to the vacuum system. The time required for extraction depends on the amount of floor affected and the amount of water present. The process is complete when liquid water is no longer being visibly collected by the vacuum system. The mats must be periodically lifted from the wood floor, and the surface must be wiped to remove the small amount of surface water. Then, the technician observes the rate at which additional water is being removed from the floor.

General rules for extraction still apply: *the more water that is extracted from under the hardwood, the quicker it will dry.*

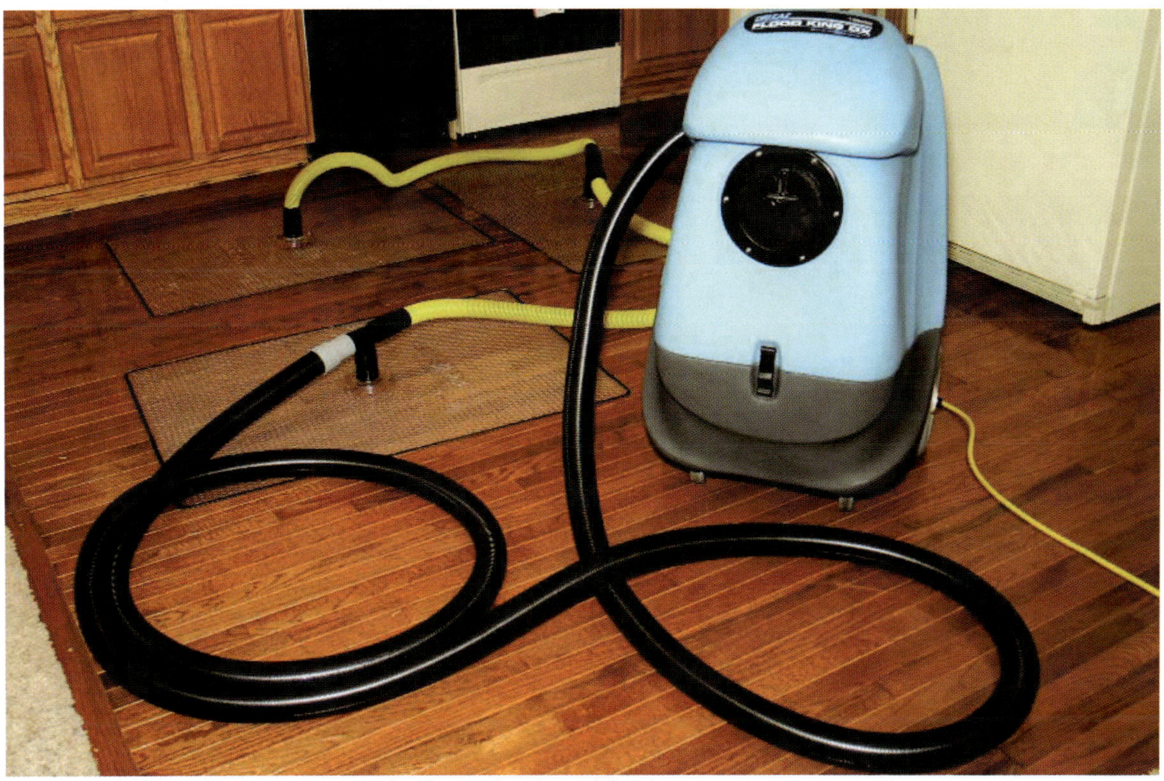

Using hardwood floor drying mats to extract liquid water from beneath flooring.

Chapter 6
Materials

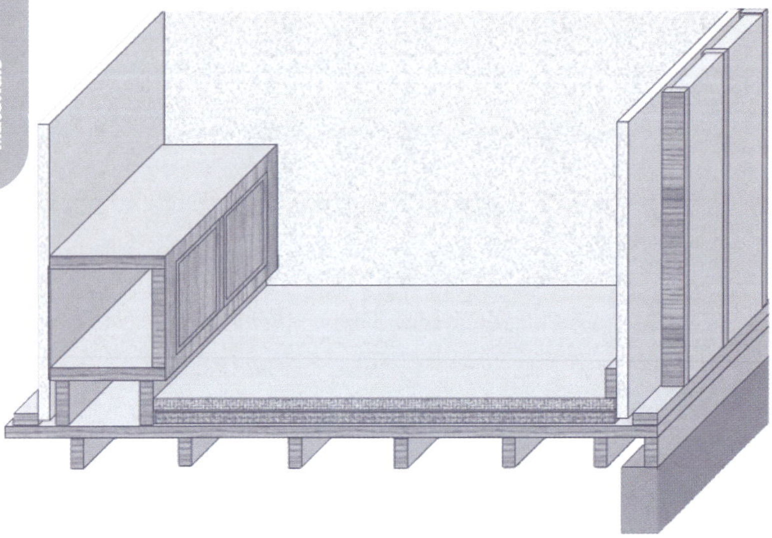

IN THIS SECTION

INTRODUCTION

Building materials used in construction today vary greatly from materials used in the past. Many of today's homes are built as "paperboard palaces," using primarily engineered materials that have been designed with cost, handling and availability constraints in mind – not durability. The fact is, many of the components used in modern building materials do not endure well when exposed to elevated moisture levels. Yet, such exposure is unavoidable.

When abnormal levels of moisture are introduced, many materials will begin to change in one way or another. Some changes – such as severe swelling and splitting – can be drastic and quite obvious. Other changes, such as hidden microbial growth, are just as serious but not as visible.

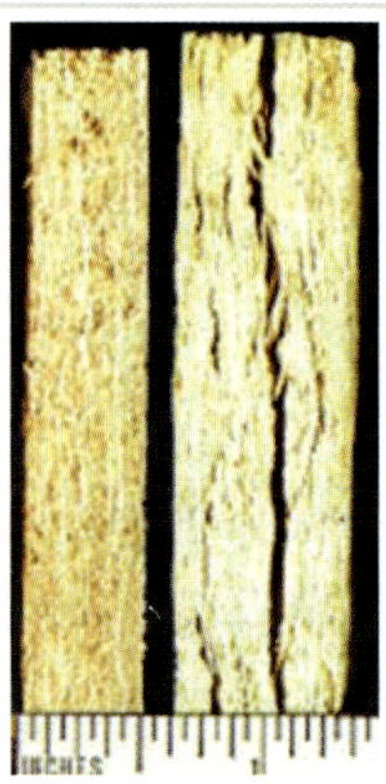

Effect of water on chipboard

As materials are dried, the nature of each specific building component and the degree of saturation affect how moisture moves through and eventually evaporates from the material. An understanding of how moisture moves through the material and what influences that movement greatly increases a restorer's success when drying affected buildings. The potential effects of abnormal moisture on common building materials vary from one material to the next – as does the way in which moisture moves through a particular material and is eventually removed.

To understand the effects of moisture on building materials, restorers should look for three forms of damage:

- **Pre-existing damage** – conditions affecting appearance or function that existed prior to the water intrusion
- **Primary damage** – effects of direct contact with a temporary water source
- **Secondary damage** – results of prolonged contact with elevated moisture levels

It is also important for restorers to understand how drying takes place. Moisture migration through materials and evaporation vary depending on the type of water present:

- **surface water** – water available at the surface of a material
- **free water** – liquid water held inside a material, deeper than the surface
- **bound water** – water absorbed and chemically bonded with the material

This chapter will first explore the three forms of damage with varying material types. The chapter will conclude with a discussion of the drying process for surface, free, and bound water.

This chapter is divided into three sections:

1 **Definitions:** This section will discuss several terms necessary to understand the varying characteristics of materials, their moisture content and their response to water intrusion.

2 **Materials** (e.g., wood, concrete): This section is divided into several parts, each focused on a specific material (and titled accordingly). Types of damage (e.g., pre-existing, primary and secondary) are discussed for each material.

3 **Drying and Materials:** This section explores how water is absorbed and released by materials, differentiating the three primary types of water encountered: surface free and bound.

DEFINITIONS

In order to fully understand the effect that moisture will have on building materials, it is important to first learn several terms and their definitions:

Damage: The physical change to a material that results in permanent deviation in appearance or function. Many materials undergo change as moisture levels increase. Some of these effects can be reversed as the material is dried. Changes that are reversible without additional treatment (i.e., refinishing) are not referred to as damage.

Pre-Existing Damage: Common physical change to material appearance, function and/or cleanliness that pre-dates the current water intrusion. Over time,

Ceiling tiles damaged by water intrusion.

building materials can be exposed to many sources of elevated moisture. These exposures cause various forms of permanent damage. It is necessary to be aware of this potential damage when each material is encountered.

Primary Damage: The consequences of direct contact with a known water source over a short period of time. Primary damage is the immediate damage to a material and is most common in engineered, cellulose-based materials.

Secondary Damage: The physical changes that occur over time if materials remain abnormally wet. This type of damage affects a wide variety of materials. Secondary damage includes dimensional changes, loss of structural integrity, microbial growth and staining. Damages caused by elevated humidity are also referred to as secondary damage. Elevated humidities are common when a water intrusion is not addressed promptly and water is allowed to evaporate without a sufficient amount of corresponding dehumidification.

Hygroscopic: Materials that are labeled hygroscopic absorb moisture easily, whether from direct contact with water or from exposure to high humidity. Such materials gain and lose moisture continually in an effort to equalize with the water content in the surrounding air mass. The more hygroscopic the material, the faster it will collect water vapor.

Again, moisture impacts specific materials in various ways – so it is important to consider each material individually. For more information on monitoring the moisture content of each material, refer to Chapter 4, specifically the section on "Measuring Moisture in Specific Materials."

MATERIALS

Wood represents the most common category of building material in many of today's structures. Wood has also undergone the most significant change over the last decade. Where solid wood was used in days past, engineered wood has been substituted. Even where solid wood is still used, the process by which it was originally dried has changed. These changes have been fueled by the need to reduce building material costs as available resources have declined.

Additionally, a desire for more resilient materials has bred the engineering and the manufacturing of "was-wood" substitutes containing a variety of additives that can complicate the monitoring and drying process.

Natural, Solid Wood

Natural wood is found as a structural member (e.g., studs, plates, joists, beams, rafters, trusses), as a finishing material (e.g., finished flooring, base trim, crown molding or coving, doorways, cabinetry) and as a component of many types of contents. Two primary types of natural wood are found in structures: *hardwoods* and *softwoods*. Hardwoods are primarily found as a finishing material, while softwoods are primarily used as structural components.

Visible staining in wooden structural members.

The metal in tack strip commonly causes visible staining if allowed to remain wet.

Pre-existing damage is found in the form of microbial growth, staining, splitting, warping and bowing of natural wood. The most evident and telling features present when solid wood has sustained pre-existing damage: *a darkening or staining from metal fasteners and staining from microbial growth.*

Chronic water problems cause severe microbial damage that manifests itself in the form of rot. In order for wood to sustain damage from wood decaying fungi (rot), moisture contents must remain above 20% MC for extended periods of time. However, damage from surface fungi occurs over shorter periods of time and at lower moisture levels.

For example, particular strains of Aspergillus and Penicillium can grow when lumber is near or above 16% moisture content. Although the growth of surface fungi does not alter the function of structural lumber, it is aesthetically damaging and its impact on human health is still being studied.

Primary damage is of little concern with natural wood. It is a hygroscopic material, but it absorbs moisture from air relatively slowly compared to other building materials. As moisture absorbs into wood, the immediate changes are minimal. When addressed promptly, the absorbed water will not create an irreversible condition (damage).

Secondary damage, however, will occur if there is a delay in responding to elevated moisture levels in and around wood. Many of the same types of damage discussed under *pre-existing damage* can occur, such as warping and microbial growth. These are all forms of permanent damage. In other words, if any of these items occur, the wooden material will require more than drying; it will need to be sanded, refinished, sealed, painted, or even replaced.

The likelihood of permanent damage is related to the way in which the lumber was milled or cut. Wood that is **quarter sawn** (hardwood) or *flat grained* (softwood) has been cut such that the annual rings create straight, parallel lines down the face of the board. Wood that has been cut in this fashion will have less dimensional change in its width as it absorbs moisture. It is more resilient to visible changes such as warping.

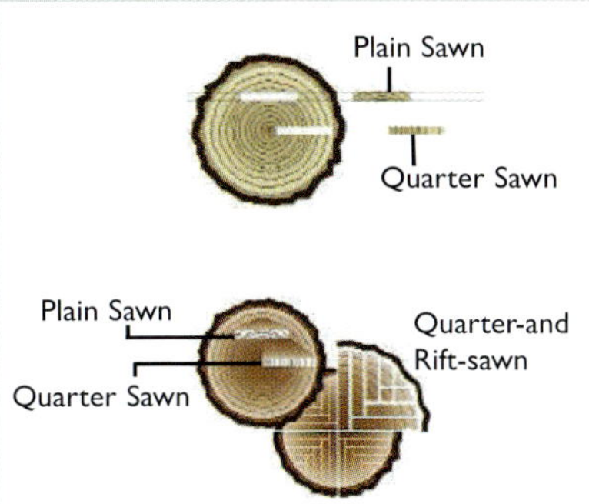

Lumber that has been **plain sawn** (hardwood) or *edge grained* (softwood) has been cut so that the annual rings create "flares" down the face of the board and form a curved grain visible on the cut-ends of the lumber. Lumber that has been cut in this way is less expensive to produce and results in a greater yield from each log. However, it absorbs and releases moisture more readily, will incur greater dimensional change, and is more likely to warp.

Quarter vs. Plain Sawn

Engineered Lumber (Was-Wood)

Many structures today are built using a variety of **manufactured wood products**. These products are used for subflooring, base and trim, floor joists, studs and even in finished flooring materials and wall paneling. The nature of each material varies depending upon the process and material used to engineer the final product. To understand the different ways in which the product will respond, it is first necessary to define the different types of materials.

The table below lists a variety of engineered wood materials and their response to wetting, then drying. When the effects of wetting and drying are considered in series, the resulting permanent damage becomes quite logical.

The more the fiber has been processed and re-adhered (laminated) in the manufacturing of the engineered product, the more rapid and dramatic the effects of wetting. **Plywood** represents the was-wood material with the least engineering, and thus the

MOISTURE AND WOOD

The materials are listed below in the order of resiliency to water damage. *Natural wood* is the most resilient to temporary water intrusion, whereas **Medium Density Fiber board (MDF)** is the least. As wood materials are engineered, they lose their natural self-bonded properties. The natural adhesion that one wood fiber has to the next is difficult to replicate. Where one wood fiber has been glued to the next, the adhesive can be jeopardized, especially when moisture is added.

	WETTING	DRYING	RESULT
Natural Wood	Minor swelling No delamination	Minor shrinkage	Full restoration
Plywood	Moderate swelling Minor delamination	Minor shrinkage	Minor delamination Minor swelling
Comply	Moderate swelling Moderate delamination	Minor shrinkage	Minor swelling Moderate delamination
O.S.B.	Moderate swelling Moderate delamination	Minor shrinkage	Minor swelling Moderate delamination
LVL Wetting	Moderate swelling Moderate delamination	Minor shrinkage	Minor swelling Moderate delamination
Particle Board	Extensive swelling Moderate delamination	Minor shrinkage	Moderate swelling Moderate delamination
M.D.F.	Extensive swelling Extensive delamination	Minor shrinkage	Moderate swelling Extensive delamination

most natural fiber bonding. It is therefore the most resilient to dramatic swelling and delamination of fibers.

Comply consists of sheets of veneer, much like plywood, on its surface. However, the core of the product consists of much smaller fibers, not veneer. Here, much artificial adhesive is used and very little of the wood fiber is naturally bonded. This results in a greater potential for delamination.

Oriented Strand Board (OSB) consists of large chips of wood laminated to form a single sheet. The amount of adhesive used is drastically greater than that of plywood, resulting in a much greater likelihood of permanent damage. This is also true of particle board. **Particle board** consists of even smaller fiber, however, and more adhesive.

Laminated Veneer Lumber (LVL) is an engineered dimensional lumber used for structural components such as studs, beams and joists. Its composition is similar to **OSB**. In the case of **LVL**, strips of natural wood fiber are laminated to form a dimensional product, as opposed to a flat sheet. Water affects **LVL** in much the same manner as **OSB**.

Finally, **Medium Density Fiberboard (MDF)** consists of the smallest natural fiber. It can be compared to thick paper sheets. Because it is the "least" natural in state, it suffers the most permanent and rapid damage.

Each of these engineered wood products can be found in a variety of areas throughout a structure. Plywood can be found as a subflooring material, sheathing, underlay and as the base for an engineered hardwood floor. Comply can be found as a subflooring material. OSB has found its way into many structural components, including engineered joists, sheathing, roofing and sheer walls. Particle board can be found as an underlayment (substrate) beneath hard surface flooring. MDF is often an inexpensive trim and baseboard material, although it can also be used as an underlayment.

In addition to these structural locations, each of the materials listed above is also commonly used for the construction of contents. Entertainment centers, shelving, cabinetry, dressers, vanities, tables, various stands and other furnishings are each made with a varying array of building materials. An important step in the inspection process is to identify each of these materials upfront. Knowing what material is used will aid in identifying **pre-existing**, **primary** and **secondary damage**.

Concrete

When evaluated as a single component, **concrete** is quite immune to water damage. Whether exposure is over a period of time, once, or intermittent, concrete will withstand elevated moisture levels. When allowed to come into contact with other materials, however, excess moisture in concrete causes ***secondary damage***. This damage includes the release or

Salt carried to the surface of concrete block by migrating water, known as effervescence.

Staining on drywall due to water damage

failure of flooring materials, swelling of wooden structural and flooring materials, staining on wood and drywall, and microbial growth on surrounding surfaces and on the soils on the surface of the concrete itself.

For these reasons it is important to inspect concrete for moisture even though the excess water will not cause damage to the concrete directly. Elevated moisture levels must be dried to an appropriate level as with any other affected building component.

Concrete is also encountered in a pre-fabricated product known as **concrete board**, also called cement board, concrete backer board, etc. These prefabricated concrete products contain no aggregate, and consist primarily of concrete and a fiber (fiberglass or **cellulose**). Concrete board is most commonly encountered as a backing material (substrate) for tile installations. Its properties are similar to concrete. It is resilient to water damage, sustaining little to no primary or secondary damage from water intrusion. The challenge it presents is in its inability to release water. It will dry relatively slowly and will therefore lengthen the amount of time that neighboring materials are exposed to excess water. Secondary damage is of concern with neighboring materials.

Gypsum

There are two materials in which gypsum can be found in structures. They are gypcrete and wall board. **Gypcrete** (leveling compound) is most commonly installed beneath tile flooring, around radiant heat systems and as a leveling material over wood and/or concrete subflooring. It is comprised of **gypsum** and sand and may also contain aggregate, wood chips, vermiculite or other materials. It is water soluble, and may show signs of pre-existing damage. Deterioration will lead to hollow sounding tile floors, dusting and cracking. In order to create these problems, contact with water must be excessive and over extended periods of time. When discovered during a temporary water intrusion, these forms of damage are pre-existing.

Primary damage is of little concern with gypcrete. Secondary damage will occur if this material allowed to remain wet. Secondary damage will include deterioration, release of finish materials, cracking and dusting. Moisture in the gypcrete will also secondarily damage finish materials such as hardwood flooring and other building materials, like wallboard, if allowed to remain wet.

Gypsum wallboard is second only to engineered wood materials (e.g., MDF) in its rapid rate of physical change when exposed to elevated moisture levels. However, gypsum has a much greater ability to recover. ***Pre-existing damage*** will be quite evident if previous exposure to moisture has occurred.

Pre-existing damage on gypsum wallboard is most evident as staining, microbial growth, swelling and change in density. Staining can be very apparent even on the painted, room-side surface of the wallboard. Microbial growth can be visible on a room-side surface and runs an even higher probability on the wall cavity interior-facing, unpainted surface. The paper layer surfacing of the gypsum is typically a cellulose material and can support microbial growth as well as, if not better than, other surfaces in the indoor environment. The paper and the gypsum are both hygroscopic; they will swell and lose much of their structural integrity when wet.

Primary damage to wallboard can be sustained if it is near complete saturation. This primary damage will occur in overhead, horizontally suspended wallboard. Wet overhead wallboard will sag – also causing a significant safety concern (see *Chapter 2 Safety*). The loss of structural integrity leaves it unable to remain flat. This sagging will result in permanent damage and requires that the wallboard be removed. Restorers must consider another safety hazard, **asbestos** and **lead** (see *Chapter 2 Safety*) when removing wallboard.

During the drying process, all of the gypsum's original strength is restored. In fact, the dried gypsum will normally be slightly stronger than *before* wetting; however, it will also be more brittle. Normally this is not a concern when dealing with drywall suspended vertically on studs. Minimal swelling will sometimes occur, causing tape joints and fastener heads to become visible. When this happens, replacement is necessary if repairs cannot be made.

Secondary damage is also possible with gypsum wallboard. Microbial growth will occur on the paper surface of the gypsum wallboard if allowed to remain wet for extended periods of time. This form of secondary damage will require removal of the affected material in most cases.

Plaster

Plaster is much more resilient than other materials used in finish construction today. When wet, it retains much of its original strength, form and function. When dried, the net result is typically quite favorable. The primary concern with plaster is its inability to release moisture in a timely fashion. Plaster can remain wet for extended periods of time because it is difficult to dry. ***Secondary damage*** is therefore the primary concern when drying plaster. When allowed to remain wet, it can affect surrounding materials – resulting in swelling, microbial growth and staining. When surrounding materials (e.g., wall studs, framing materials) are allowed to swell, secondary damage to plaster can take the form of cracking.

Brick

When evaluated as a single component, **brick** is quite immune to water damage. Whether exposure is over a period of time, once, or even intermittent, brick will withstand the effects of elevated moisture levels. But when it is allowed to come into contact with other materials, brick with an elevated moisture level can cause ***secondary damage***. This damage can include the release or failure of finish materials, swelling of wooden structural and flooring materials, staining on wood and wall board, and microbial growth on surrounding surfaces and on soils on the surface of the brick itself. For these reasons, it is important to inspect brick for moisture even though the excess water will not cause direct damage. Brick must be dried to an appropriate level – as with any other building component.

Carpet

A variety of construction methods and materials are used in **carpet** today. In order to assess pre-existing conditions, it is first important to identify the type of carpet affected – including the way it was made and the materials used. Below are the three most common construction methods found in U.S. structures:

Cut pile. Nylon and **polypropylene** fibers are used to make the most common type of residential carpet: **cut pile**. The fibers are tufted into a polypropylene mesh backing (primary backing) and adhered using a latex adhesive. When manufactured, the face fiber is a continuous yarn. After being tufted into the backing, the yarn is cut, and therefore named a *cut pile.* The primary backing is then laminated or adhered to a secondary backing, using latex adhesive. This secondary backing serves as a structural reinforcement. Nylon and polypropylene are not hygroscopic and are resilient to water damage. The latex adhesive is hygroscopic and will lose as much as 50 to 80% of its strength when wet.

Berber. The second most common carpet in residential construction is the **berber** carpet, which is constructed similarly to cut pile. The most significant variance with Berber carpet is that the yarn, when tufted into the primary backing, is left intact. The continuous yarn then forms a loop pile where no end of the fiber is visible. Another deviation from cut pile carpets is the choice of fiber. Normally, an **olefin** (polypropylene) fiber is used as opposed to **nylon** fiber. Olefin is not a **hygroscopic** material.

Woven. The least commonly found carpet in residential construction is the woven carpet. Woven carpets are constructed exactly as the name indicates. Yarns are woven through a backing material to form a thick, dense pattern of face fiber. The fibers used in woven carpets vary greatly – from synthetic materials such as nylon and olefin to wool, cotton and other natural materials.

The natural materials used in many woven carpets are hygroscopic and will typically swell when absorbing moisture. Dying natural fibers is typically a process using moisture, where natural fibers absorb the moisture and the dye it carries. This leaves some woven carpets vulnerable to dye transfer when wet.

In addition to carpet construction, the method of installation varies. There are two primary methods for carpet installation:

Stretched-in carpet is the most common method of installation found in residential structures. Stretched-in carpets are typically installed over a carpet **underlay** (pad, cushion). Around the perimeter of the room, a wooden strip is installed that contains several tacks angled slightly toward the wall. The carpet is installed by forcing it toward the walls to form a tight, taut and secure stretch. The carpet is held in place as the **backing** is placed over the multiple tacks. This process is completed using specialized tools – a **power stretcher** and a **knee kicker**.

Direct glue carpet is most commonly found in commercial structures. "Double stick" methods are also used when an underlay is glued to the subfloor, then the carpet is in turn glued to the underlay. Glued down carpets are most commonly glued directly to the subfloor. Many glue-down carpets consist of either polypropylene or olefin face fiber and are constructed as a single yarn, left intact – adhered to a much more dense backing than cut pile or berber carpets. Backing materials are often made of rubber or other dense closed cell foam. The rubber and foam backings used are not normally hygroscopic, nor are the face fibers commonly used.

Once the carpet type has been identified, it can be evaluated for ***pre-existing conditions***. Stretched-in carpet installations can show signs of many pre-existing conditions. These may be the result of previous water intrusions, but they can also be the result of improper installation and handling; insufficient care and maintenance; or improperly specified underlay.

Pre-existing conditions found in stretched-in carpet include:

- Separating seams
- Rippling appearance
- **Delamination**
- Fraying or damage to fibers
- Loss of fibers or "balding" of fibers
- Staining and microbial contamination

Organic debris will be present because carpets act as the largest filter in the indoor environment, trapping many particulates, soils and dusts. If not properly cared for, this buildup of organic material will support a rapid deterioration of cleanliness when in contact with excess moisture. Additionally, urine may be present from pets. Other forms of biological waste may be present depending on the structure's type and normal use.

Damage to fibers and a separation of the primary and secondary backings (delamination) occurs when traffic lanes have been in the same area of the carpet over a number of years. Delamination also occurs if the carpet is not properly installed and does not have the proper tension associated with power stretching. Another common cause of delamination is the use of improper carpet underlay. Carpets that have been installed over underlay (cushion, pad) that is too soft will have excessive movement in the carpet backing, leading to deterioration of the latex adhesive over time.

Carpets comprised of natural fibers (e.g., cotton, wool) will suffer primary and secondary damage. These natural fibers are hygroscopic, and will absorb moisture. As the fibers absorb moisture, they swell. This creates a thicker yet shorter yarn, causing shrinkage.

The most common carpets used in structures today consist of synthetic fibers and will not suffer directly from water intrusion. ***Primary damage*** typically will not occur if the excess moisture is addressed correctly. While wet, carpet will lose as much as 80% of its strength as the latex adhesive softens. If wet carpet is handled properly, there should not be permanent damage to the carpet.

The extraction and drying processes can be laborious, however, and may exceed the cost of the carpet replacement. If pre-existing conditions are present, carpet may need to be removed. The excess moisture and the physical aspects of extraction may worsen the condition.

When allowed to remain wet, carpet can suffer from ***secondary damage***. This damage will occur in the form of microbial growth, color transfer, staining and a loss of structural integrity. Although most carpeting contains mainly synthetic materials, the organic debris present (discussed earlier) can support an amplification of microorganisms.

Contents

Understanding the influence moisture has on **contents** begins with material identification. Materials such as engineered wood, natural wood, and fabrics similar to those in carpet fiber, are all common in building contents – and all warrant further discussion.

There are specific areas of concern when evaluating potential damage to contents. ***Pre-existing conditions*** may exist in the form of scratches, wear, cracks, warping and other signs of use, mishandling and poor maintenance. When drying a water damaged structure, contents will need to be handled and moved by restoration technicians. When pre-existing damage is present, it is imperative to document. If this damage is not documented, the restorer may be blamed for causing the scratch, crack or warping.

Many contents are at risk for ***primary damage***. A number of **cellulose**-based materials are found in contents ranging from papers and cardboards to MDF and particle board. When wet, these highly **hygroscopic** materials will rapidly swell and, in most cases, will become permanently damaged.

Secondary damage can occur in every form previously mentioned for each material. Specific types of secondary damage associated with contents are typically:

- **color transfer**
- **swelling**
- **delamination of surface/finish material**
- **staining**

The dying processes used for the variety of rugs, sheets, clothing, drapery and furniture fabric are not always colorfast. This can lead to rapid transfer of color. In addition, most contents are capable of transferring tannin from wooden legs to the wet car-

Although wood may appear dry, it always contains some moisture.

pet. Rust can also eventually stain the carpeting below when metal legs are not separated from the moisture.

DRYING AND MATERIALS

The rate at which moisture moves through materials depends upon two primary factors: the type of material affected and the degree of wetness. The type of material affected will vary in permeability (ability for water to pass through the material), hygroscopicity (ability to absorb water), thickness, density, temperature and **"R" factor** (**resistance to heat transfer**).

Each of these characteristics will influence the way in which water travels through the material. The degree of wetness will influence moisture movement, in combination with other material characteristics, by indicating the type of water being addressed. Water will be present in one, two or three forms:

Surface water is readily available at the surface of a material. It is liquid, visible and will evaporate readily. All materials are capable of supporting surface water. Surface water is best removed physically using mechanical extraction equipment or other physical means. Minute amounts of remaining surface water are readily evaporated using large amounts of airflow and moderate temperature and humidity.

Due to the ease of access to surface water, extremely warm temperatures and low humidity are not necessary for a sufficient rate of evaporation. Surface water yields the fastest evaporation rate.

Free water is present within a material but is not bonded. It exists in cavities, open pores and other air spaces within the material. It is liquid, generally visible by a darkening of the material (e.g., wood) and can evaporate readily. Most materials are capable of supporting (containing) free water. Materials not capable of retaining measurable amounts of free water are non-permeable, non-porous materials such as vinyl, **vinyl composition**, steel, rubber and some other solid, synthetic materials.

Free water is best addressed with sufficient airflow, warm temperatures and moderately low humidity. Because free water is not readily available at the surface of the material, lower humidities and warmer temperatures are more necessary than with surface water.

Free water is evident in wood when it appears dark, or saturated.

As the free water nearest the surface of the material is depleted, lower humidities and warmer material temperatures are necessary in order to maintain the rate of evaporation. For a more in-depth discussion of this process, refer to *Chapter 7 Drying Science* and *Chapter 9 Drying Systems.* Free water will evaporate quickly.

Bound water is absorbed by the material and held captive by chemical bonds. It is similar to vapor in that it is not fluid, yet similar to a solid/liquid in that it is bonded to other molecules. Because it is bound to other molecules, it must first be freed before it can be removed.

Airflow is least effective when trying to remove bound water from within a material. Because the moisture must first be freed (evaporated) before it can begin moving toward the surface (and eventually into the surrounding air mass), materials will dry fastest if they are warm and the surrounding humidity is very low. The warm temperature and low humidity are discussed in detail in *Chapter 7 Drying Science.* Bound water evaporates very slowly in comparison to free and surface water.

Wet indoor environments are an accumulation of varying building materials at varying degrees of saturation. Evaluating materials to understand what form of water they support will yield an understanding of how quickly moisture will evaporate, what rate of airflow is necessary across wet surfaces, how much heat will be required, and how low humidity must be in order to facilitate an acceptable rate of evaporation.

If surfaces are properly evaluated, progress will be most evident in the initial stages of the drying process. In fact, surface water should be eliminated in the first several hours. After that, free water will begin to evaporate from the most porous and permeable materials (e.g., carpet, wall board with permeable paint). Materials that are retaining bound moisture will remain wet or possibly increase in moisture content during this stage.

As free water begins to dissipate, humidity will begin to drop. The rate of evaporation will slow dramatically – allowing for humidity control to yield a net reduction in overall atmospheric moisture content. As the humidity continues to drop, bound water will eventually respond. This process can be sped up with the appropriate balance of temperature applied to affected materials.

An increase in material temperature will help maintain the initial rate of evaporation, as warmer materials will most directly affect bound water. The process of applying heat to bound water counteracts the chemical bond the water has with the material. Several limitations and complexities exist when increasing material temperatures. Restorers must be aware of these problems in order to prevent severe secondary damage. More information on this topic is found in *Chapter 7 Drying Science* and *Chapter 9 Drying Systems.*

Chapter 7

Drying Science

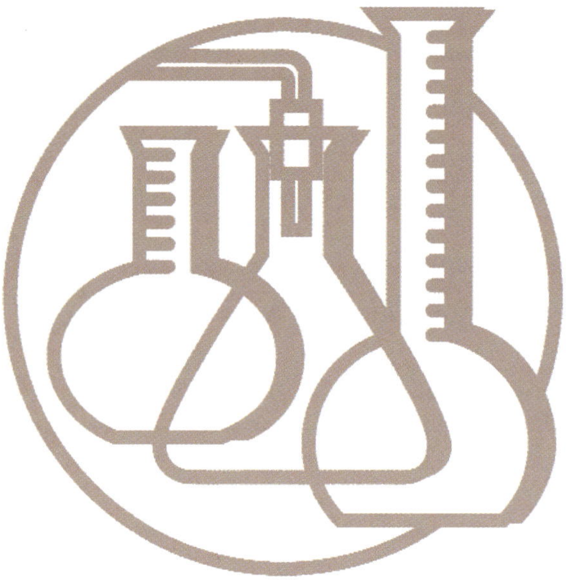

$$Ft^3 \times 7.48 = g$$

$$\frac{RH}{100} = AW$$

$$\frac{Lbs./Day \times GD}{7{,}000} = Pints$$

$$\frac{CFM \times 60 \times 24}{14} = Lbs./Day$$

IN THIS SECTION

INTRODUCTION

Drying structural materials requires an understanding of drying science and the laws of physics. This chapter will explore the application of these natural laws and science, beginning with one of the most important: psychrometry.

PSYCHROMETRY

What exactly is psychrometry? *"Psychrometrics deals with thermodynamic properties of moist air and uses these properties to analyze conditions and processes involving moist air." (1997 ASHRAE Handbook, 6.1 p1)*

In simpler terms, psychrometry is the science and practice associated with measuring the properties of air, namely temperature and humidity, and analyzing these properties to better understand their effects on the surrounding environment.

There are three basic properties evaluated in psychrometrics that this chapter will discuss. These are **Humidity**, **Airflow** and **Temperature** (HAT). Humidity, airflow and temperature are commonly

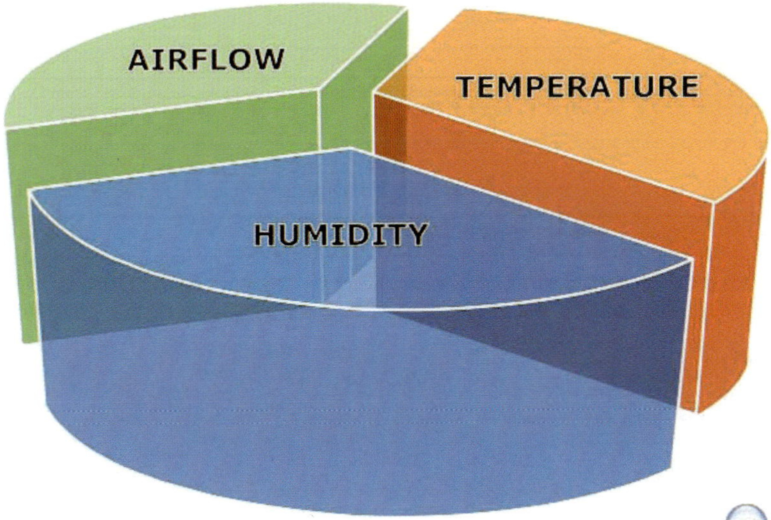

Greater rates of evaporation will occur when liquid water is made warmer and when drier air is moved more rapidly across the wet surface.

referred to as the drying pie and abbreviated as H.A.T. Each can be measured and evaluated in a number of ways. Each has a direct influence on the rate of evaporation (or condensation) that will occur when moisture and materials interact.

A common misconception among restorers is that either evaporation or condensation occurs, depending upon temperature; where surfaces are cool, condensation will occur, and where surfaces are warm, evaporation will occur. In reality, condensation and evaporation are both constant. Every body of water is in a constant battle for survival, continually losing and gaining moisture, to and from the surrounding air.

Two major factors determine which will occur at a greater rate: the water's temperature and the air's moisture content (humidity). In order for evaporation to win the battle and for the body of water to have a net "loss," the water must have enough energy to promote a sufficient rate of evaporation. To deposit moisture at a slower rate, air must not contain high levels of humidity.

While water temperature and air moisture content define whether condensation or evaporation will win the battle, *airflow* determines the rate at which the battle will play out. Where no airflow exists, the net change will be slow. Where airflow does exist, the net change will occur at a more rapid rate.

Humidity, airflow and temperature directly affect the state in which water exists – **solid** (ice), **liquid** (water) or **gas** (vapor) and the rate at which the change of state occurs. It is critical to have a clear understanding of the methods for measuring psychrometric properties and of how to use them appropriately when evaluating a restorative drying environment.

Ultimately, the goal in restorative drying is to remove excess water from affected building materials after a water intrusion. This process requires that restorers change liquid water into vapor (evaporation). Once water is evaporated, it must then be removed. This is most often done by changing vapor (humidity) back into a liquid (condensation) using refrigerant dehumidifiers. Therefore, the goal in restorative drying involves *changing water s state.*

THE INFLUENCE OF TEMPERATURE

Water exists in three states: **solid** (ice), **liquid** (water) and **gas** (vapor). The primary factor that will ultimately determine what state water will take is the amount of *energy* it contains. The more energy each molecule possesses, the more rapidly it moves. When molecules are moving quickly enough, the chemical attraction they have to each other is no longer sufficient to hold them together. As water molecules gain or lose enough energy to create or break bonds with one another, change of state occurs.

There are several changes of state that can occur, depending upon whether energy is being added or removed. The amount of energy required to change water's state is quite phenomenal. In fact, it requires more energy to change water from one state to the next than is required with almost any other molecule.

Energy is measured in the restorative drying industry using dry bulb temperature, or simply degrees

Solid

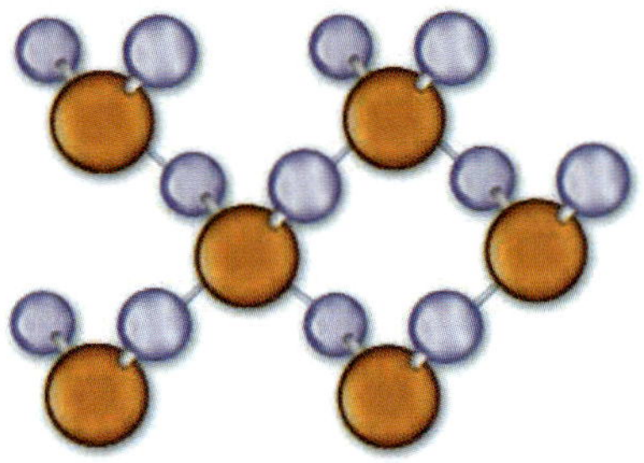

Liquid

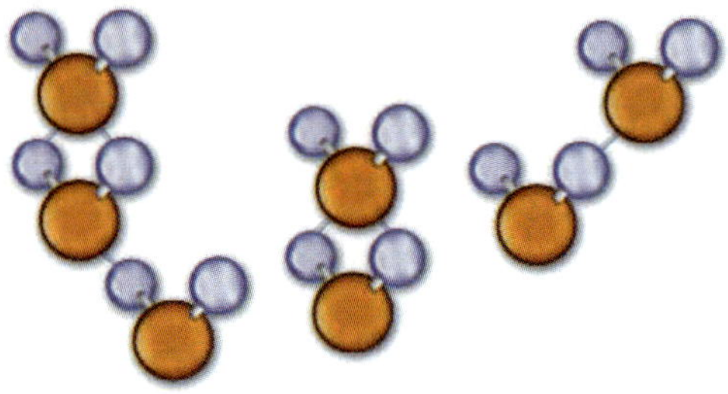

Gas

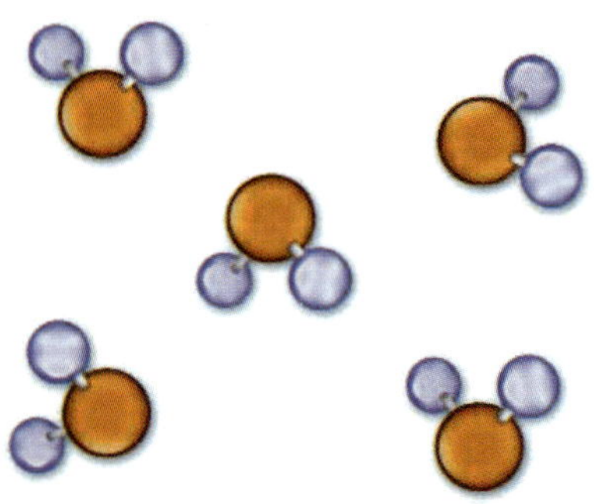

Fahrenheit (Celsius). Where temperature is higher, more energy is present. It is important to note that temperature measurement is valuable, but is not a specific measurement of heat. Heat energy is measured by other means, commonly in British Thermal Units, or BTUs (United States, Canada) and in Calories (Metric). *Chapter 8 Drying Equipment* will discuss how to use BTUs (Kilowatts) to control heat produced by drying equipment.

Energy Required to Change State

Energy is measured in British Thermal Units (BTU or Kw). A BTU is the amount of energy necessary to raise one pound of water one degree Fahrenheit. (To convert BTUs into Kw, divide BTUs by 3412.) Changes in state, and the amount of energy required for each, are:

- **Melting:** Solid changing to liquid. Occurs as energy is increased. The amount of heat necessary to change one pound of solid water to liquid water is 143.4 BTUs.
- **Freezing:** Liquid changing to Solid. Occurs as energy is removed. Requires that 143.4 BTUs are removed from every pound of water.
- **Evaporating:** Liquid changing to a vapor. Occurs as energy is increased. The amount of energy necessary to change one pound of liquid water to vapor is 940 BTUs.
- **Condensing:** Vapor changing to a liquid. Occurs as energy is removed. Requires that 940 BTUs are removed from every pound of water.

This information is not commonly used by restoration professionals when making drying decisions. The knowledge is valuable, however, when considering the amount of effort necessary to evaporate liquid water. As discussed in *Chapter 5 Extraction*, physical water removal is critical to restorative drying success. Restorers must remove as much physical water as is possible, using evaporation to address what cannot be physically removed.

THE INFLUENCE OF HUMIDITY

Humidity plays an important role in restorative drying. The amount of moisture in air will directly affect the amount of moisture in materials. Humidity also influences the rate at which materials will gain or lose their moisture. The relationship between moisture in air (humidity) and moisture in materials is both relative and specific.

Relative Equilibrium

Phase 1: Relative imbalance

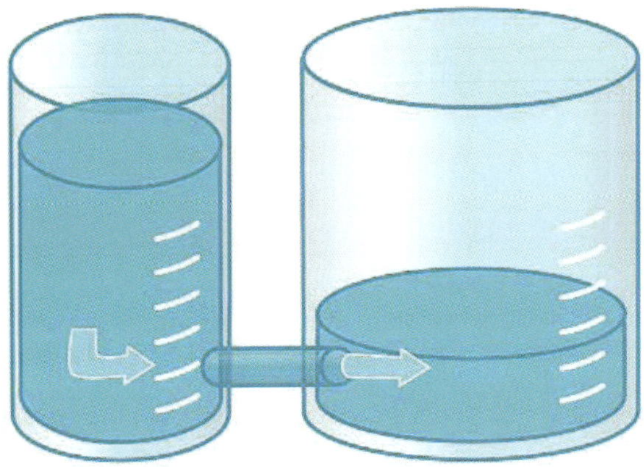

Phase 2: Relative equilibrium

Phase one shows the container on the left with a relatively higher water level than the container on the right. Phase two shows *the result of relative equilibrium*.

The Relative Relationship

The amount of moisture in materials is most directly related to the **relative humidity (RH)** of the surrounding air. RH is the amount of moisture in air expressed as a percentage of the air's total holding capacity. It is an expression of how much water the air is holding compared to what it *could* hold.

Relative humidity can be compared to a fuel gauge in an automobile. The gauge displays how much fuel is left in the tank using marks that denote percentage. The tank may be half full, for example, which would represent 50%. If the tank is 50% full, the driver understands that the tank contains only half of what it could hold. The driver does not know, however, how much fuel is actually in the tank. This would require a measurement in gallons (liters) or pounds (grams).

With the measurement of relative humidity, restorers have information similar to the driver's fuel gauge. Air that is at 50% RH contains half of the water it could hold. The restorer knows only that the air is half full, and does not know how much water is actually present. The amount of moisture air can hold changes with temperature. When air is warmer, it can hold more moisture.

Changes in temperature will therefore change the relative humidity. Because the capacity for moisture increases as temperature increases, the same amount of moisture will represent a smaller percentage of the total capacity. Relative humidity will decrease as air temperature is increased. The opposite is also true. When air temperature decreases, its capacity for water vapor also decreases. The same amount of water will now represent a greater percentage of the total capacity. As air temperature decreases, relative humidity increases.

Materials will seek equilibrium with the surrounding relative humidity. For example, if an environment was to become more humid, and the relative humidity increased to 70%, **hygroscopic** materials would absorb moisture until they were holding approximately 70% of their holding capacity. This is referred to as a relative equilibrium.

The Specific Relationship

The rate at which moisture will move is most related to the actual amount of water in air. In addition to relative humidity, humidity can be measured as a weight. This measurement is much more specific to the amount of water in air. It also directly affects how quickly water will move. Where water vapor in air weighs more, it exerts more pressure. This pressure pushes back against water leaving materials, or adds force to water entering materials.

Specific Equilibrium

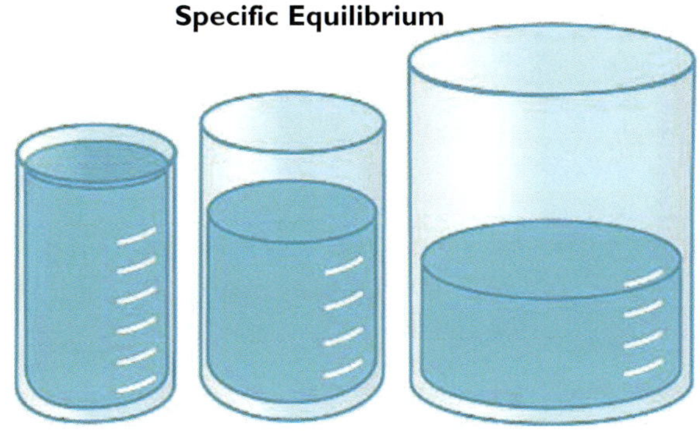

Although the amounts of water in these three glasses are the same, each contains a different amount relative to its capacity.

The weight measurement used in the restorative drying industry (United States, Canada) is **grains per pound** of dry air, abbreviated as **GPP**. It is a ratio, expressing the weight of water in a given weight of air (a pound of air). The metric equivalent to this measurement is grams per kilogram of dry air, abbreviated as **g/kg**.

GPP (g/kg)

The measurement of GPP (g/kg) is a measurement of humidity ratio. The amount of water in air is expressed in weight within a given weight of air. GPP, used in the United States and Canada, uses grains to express the water weight in a pound of air. Grains come from the agricultural industry and are exactly what the word indicates: grains. A pound of wheat contains approximately 7000 individual grains. This measurement of weight has found its way into several industries and is used to measure several things, such as pharmaceuticals, gun powder, and even water vapor. A measurement of GPP is therefore the grains of water vapor (literally, 1/7000 pound) in every pound of air. It is important to note that a pound of air is approximately $14ft^3$, or the size of a four drawer filing cabinet.

The metric measurement is grams per kilogram of dry air. Water vapor weight is expressed as: grams of water vapor in every kilogram of air (g/kg).

RELATIVE VS. SPECIFIC

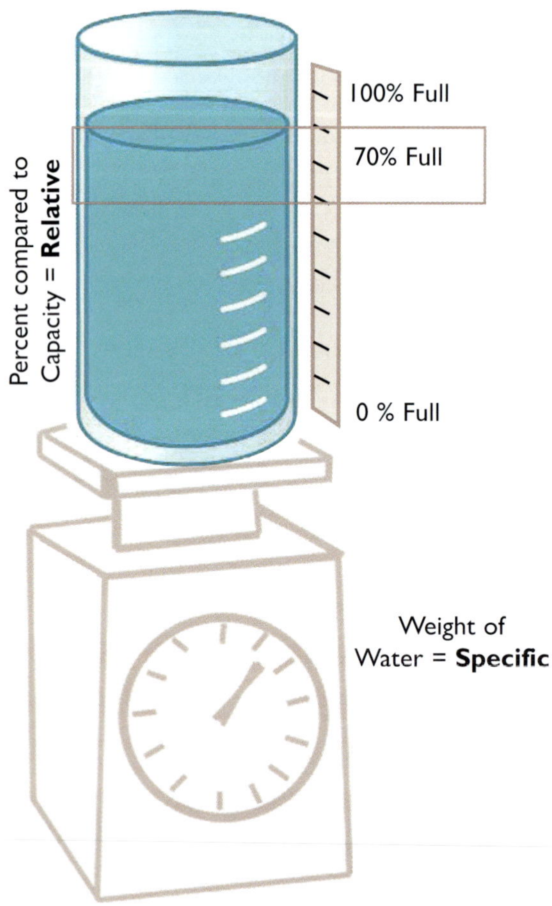

Because this measurement is of the water vapor weight, it tells the restorer where the moisture is in greatest concentration. Greater weight means more moisture. Higher GPP (g/kg) in the air surrounding a material will retard the evaporation process because the weight of water vapor resists vapor leaving the material. GPP (g/kg) therefore influences the rate at which moisture moves toward equilibrium.

THE INFLUENCE OF AIRFLOW

Air movement in the drying environment speeds the effect of humidity and temperature. It provides a blending effect, ensuring that the warm, dry air created by drying equipment (e.g., dehumidifiers) is quickly and continuously placed into contact with wet materials. It also ensures that evaporated water vapor is removed from these same surfaces.

Another important purpose of air movement is to create air exchange between adjacent affected areas. Without airflow, humidity and temperature would greatly vary from one area to the next. Restorers would need to install drying equipment in every affected air space within the structure. This would be impractical for most applications.

It is necessary to consider the need for this air circulation throughout all affected areas when installing air movers. Installation of drying equipment is discussed in *Chapter 8 Drying Equipment*.

The more moisture a material contains, the faster water will evaporate. Greater evaporation rates require more airflow to maintain low humidity along the surface. As materials dry, less airflow is necessary. This relationship is directly related to the type of water present. Where surface water is present, the greatest amount of airflow is necessary. Where only bound water is present, the least amount of airflow is necessary.

THE PSYCHROMETRIC CHART

To properly evaluate temperature, relative humidity and humidity ratio (GPP, g/kg), restorers must use a psychrometric chart. Other tools are available to calculate GPP (g/kg) and can be useful (e.g., electronic psychrometric calculators). Typically, a restorative drying technician will acquire a dry bulb temperature (Fahrenheit or Celsius) and a relative humidity using a thermo-hygrometer. These two readings establish the size of the air mass (temperature) and the relative amount of moisture in the air (relative humidity).

Using these two values, the technician will use the psychrometric chart to calculate GPP (g/kg). Other calculations are also made if necessary, including vapor pressure, dew point temperature and thirst (discussed later in this chapter). Once these readings are established, the technician compares them to surrounding conditions or previous conditions to evaluate drying potential.

Drying potential is evaluated on the psychrometric chart in a number of ways. Using the temperature, relative humidity and GPP (g/kg), the restorer can:

- Determine if outdoor air can be used to promote drying. This is done by comparing outdoor GPP (g/kg) to indoor. Outdoor air is only used to support drying when the GPP (g/kg) outside is lower. If this is not true, then outdoor air is wetter than indoor air and is not used.
- Determine the effects of heating or cooling an air mass. This is necessary when using outdoor air. Typically, the temperature of outdoor air is manipulated (heated) as it is brought into the structure. Technicians can first calculate the outdoor GPP (g/kg), then determine the new indoor relative humidity. When the outdoor air is heated, the GPP (g/kg) will remain the same, and relative humidity will decrease.
- Evaluate the drying system/control. This is done by comparing current GPP (g/kg) in the affected area to previous days. The GPP (g/kg) will decrease over time. If it does not, then evaporation and dehumidification are not appropriately balanced and a change is necessary to equipment configuration.

PSYCHROMETRIC CHART

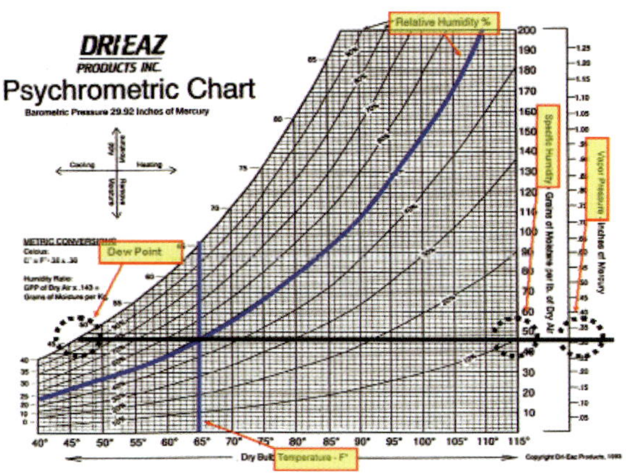

Many calculations are made using a psychrometric chart.

OTHER CALCULATIONS & MEASUREMENTS

The most basic application of psychrometry is the use of temperature and relative humidity to calculate GPP (g/kg) as discussed earlier in this chapter. In addition to these, however, there are several other important calculations and measurements restorers must consider. These are:

- **Dew Point Temperature:** The temperature at which air becomes saturated
- **Vapor Pressure:** The force exerted by water vapor on the surrounding environment
- **EMC & ERH:** Equilibrium Moisture Content (EMC) and Equilibrium Relative Humidity (ERH)
- **Thirst:** The capacity in air for additional water vapor
- **Permeance:** The ability for water vapor to travel through a material
- **Grain Depression:** Water removal calculation for dehumidification equipment

The following sections will discuss each of these items and their use in the restorative drying environment.

DEW POINT TEMPERATURE

Dew point is the temperature at which air becomes saturated, or can hold no more water vapor. When air is cooled, its capacity for water vapor is reduced.

Eventually, if temperature is cooled enough, the relative humidity will reach 100%, or saturation. When this occurs, the air has reached its dew point temperature.

Calculating the dew point temperature is especially critical when monitoring humid air masses. Typically, the most humid air mass in a structure affected by water intrusion will be the indoor air. However, at times the most humid air mass will be the outdoor air. It is necessary to calculate the dew point temperature for both of these air masses. Restorers then compare the dew point to surface temperatures throughout the structure.

If surfaces are cooler than the surrounding air, they will affect the air temperature. In fact, if a surface is cool enough, it will dramatically reduce the air temperature. Because reducing air temperature will increase relative humidity, this can result in the air reaching dew point temperature, or saturation. As discussed earlier, the relative humidity of air plays a critical role in evaporation. If humidity is high, evaporation will be retarded. It is possible that the process of condensation may even overcome the evaporation rate, causing materials to gain moisture as opposed to losing it. Evaluation of dew point is therefore a critical component in understanding evaporation potential.

tip

Dew point temperature is typically calculated by using the psychrometric chart in the following way:

- Obtain the ambient dry bulb temperature (Fahrenheit or Celsius) and the relative humidity (using a thermo-hygrometer).
- Plot these figures on the psychrometric chart.
- Where the temperature line intersects the relative humidity line, draw a horizontal line.
- This line will intersect with the dew point temperature at the far left of the chart.

The line that reflects the dew point temperature is actually the 100% relative humidity line. Each curved line on the chart represents a relative humidity, drawn in 10% increments. The bottom edge of the chart represents 0% relative humidity, the first curved line represents 10%, the second represents 20% and so on.

Observe the effects of cooling air by following the horizontal line drawn through the intersection of temperature and relative humidity. As the line is drawn to the left, it crosses several of the curved relative humidity lines.

In effect, as the line moves to the left, air is becoming cooler and cooler. It is important to note the temperature immediately beneath this line as it is drawn to the left. The further the line is drawn to the left, the cooler the dry bulb temperature.

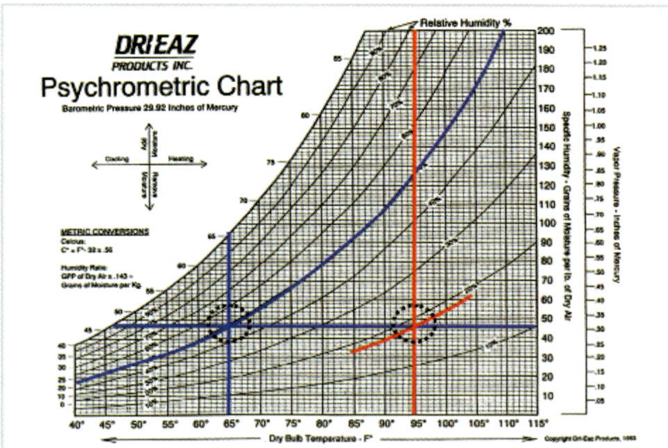

Changing the temperature of air will affect relative humidity. The actual amount of water in the air (GPP) does not change as the temperature is lowered. The relative humidity, however, increases.

VAPOR PRESSURE

Vapor pressure is *the force exerted by vapor on the surrounding environment.* This pressure is directly related to the vapor's weight (GPP, g/kg). Vapor pressure is calculated using the psychrometric chart, similarly to the way dew point is calculated. Vapor pressure is the last unit of measure on the far *right* hand side of the chart.

The scale for vapor pressure is immediately to the right of the scale for GPP (g/kg). This is important because higher and lower water vapor content is measured and evaluated using GPP (g/kg). The greater the weight of water vapor, the greater the vapor pressure; therefore, vapor pressure and GPP (g/kg) are directly related to each other.

Water vapor pressure is the force that drives higher concentrations of humidity from one space to another – in an attempt to seek equilibrium with lower concentrations of water vapor. Vapor pressure is not only responsible for water's equilibrium between one air mass and another, but it is also responsible for the rate of equalization between moisture in materials and moisture in the surrounding air mass.

EQUILIBRIUM

As discussed earlier, the vapor pressure in air influences the direction in which water vapor will move. Pressure will always move from highest to lowest. This pressure is calculated using a psychrometric chart. The calculation for vapor pressure in air has already been discussed. Measuring this pressure associated with moisture in materials, however, requires additional information.

Understanding the relationship between air vapor pressure and material vapor pressure requires knowledge of two additional terms: Equilibrium Moisture Content and Equilibrium Relative Humidity.

The amount of moisture in a material and the relative humidity of the air immediately surrounding it have a very direct relationship to each other. In fact, the relationship is *so* direct that the amount of moisture in a material can be measured using a hygrometer. By containing a sample of air at the surface of a material (or within the material) and allowing that air to achieve an equilibrium with the moisture content of the material itself, the relative degree of saturation can be measured. A material holding very little water compared to its "saturation" will support a very low relative humidity. Conversely, a material near saturation point will support very high relative humidity.

Equilibrium Moisture Content

RH	30° F	40° F	50° F	60° F	70° F	80° F	90°F
20%	4.6	4.6	4.6	4.6	4.5	4.4	4.3
30%	6.3	6.3	6.3	6.2	6.2	6.1	5.9
40%	7.9	7.9	7.9	7.8	7.7	7.6	7.4
50%	9.5	9.5	9.5	9.4	9.2	9.1	8.9
60%	11.3	11.3	11.2	11.1	11.0	10.8	10.5
70%	13.5	13.5	13.4	13.3	13.1	12.9	12.6
80%	16.5	16.5	16.4	16.2	16.0	15.7	15.4
90%	21.0	21.0	20.9	20.7	20.5	20.2	19.8

For Oak Species Hardwood – Moisture Contents are approximate and subject to variation.

This relationship is best understood using an Acclimatization Chart or Equilibrium Moisture Content (EMC) chart. An EMC chart shows the relationship between moisture content and the psychrometric properties of air. Although temperature, GPP (g/kg), vapor pressure and relative humidity all have an effect on the amount of moisture in a material, relative humidity is the most closely related. *An increase in relative humidity will result in an increase in moisture content.*

This fact can also be stated in the reverse: *An increase in moisture content will result in an increase in relative humidity, if the air at the surface of the material were allowed to acclimate with it.* This relates to vapor pressure because when the humidity is measured at the surface of the material (or internally), the humidity created by the moisture level is known.

This humidity value, referred to as an **Equilibrium Relative Humidity (ERH),** can be placed on a psychrometric chart and used to calculate the vapor pressure associated with the moisture in the material. Once the ERH has been plotted on the chart, the material's temperature must be measured, and then plotted as the dry bulb temperature. Where temperature and ERH intersect, a horizontal line can be drawn to the right to calculate the material vapor pressure.

The restorer can then compare the material vapor pressure to the vapor pressure associated with the surrounding air. If the vapor pressure in the air mass is greater than the material vapor pressure, drying will not occur. There are two options for solving this problem: either *increase the material vapor pressure* or *decrease the vapor pressure of the air.*

Material vapor pressure can be increased by raising the material's temperature. The effect of raising temperature is calculated using a psychrometric chart. The new, higher temperature is drawn to intersect with the ERH line obtained from the initial surface (or internal) humidity reading. This ERH does not change because it is directly related to the amount of moisture in the material. It will not decrease until the material becomes drier. The new vapor pressure is determined by drawing a horizontal line to the right where temperature and relative humidity intersect.

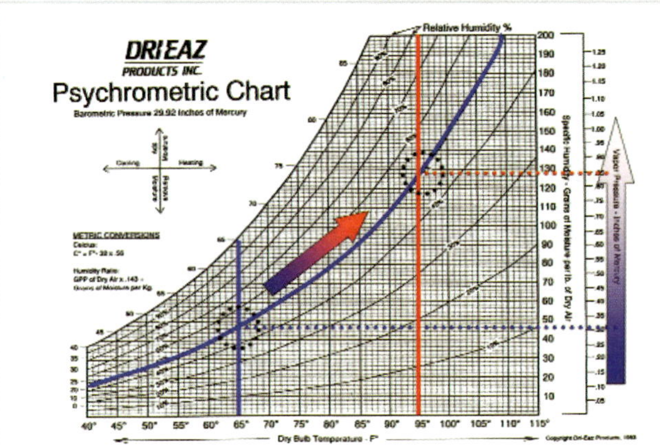

Material vapor pressure can be increased by heating the wet material. The material's vapor pressure increases, in this case from approximately 0.31 in. of mercury to 0.85 in. of mercury.

Decreasing the air's vapor pressure requires dehumidification. Lowering the GPP (g/kg) of the air will decrease the vapor pressure of the air because they are directly related. Traditionally, the air vapor pressure is the focus of restorative drying because lower humidity prevents many forms of secondary damage.

Vapor pressures of both materials and of air must be evaluated, however, to maximize the rate of evaporation. If material vapor pressure is lower than air vapor pressure, drying can not occur. This will be common when materials are cool, when trying to dry materials that are already relatively dry or when humidity in the drying environment is relatively high.

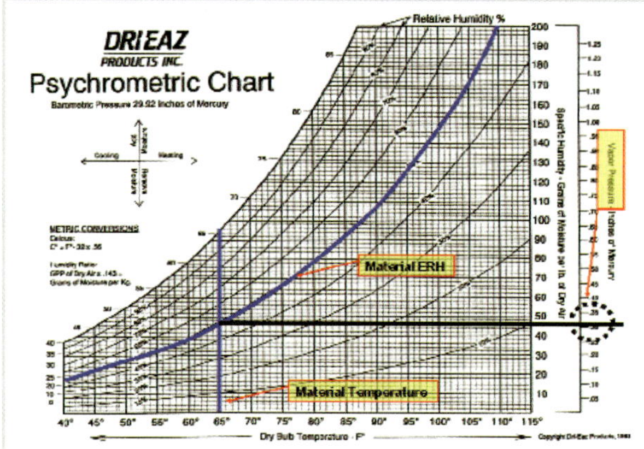

Calculating material vapor pressure

THIRST

Evaporation rates increase when materials are made wetter and relatively warmer; when air is made relatively drier; and when more airflow can place drier air and more energy in contact with the wet surface. The calculations discussed to this point each look at one of the influences on evaporation. Thirst, however, is *a psychrometric calculation that considers both the amount of energy and the amount of moisture present in air.* Thirst is calculated by plotting relative humidity and temperature on a psychrometric chart, then comparing the GPP (g/kg) to the total capacity the air has for water vapor at the same temperature.

When attempting to promote a greater rate of evaporation, technicians must evaluate the thirst of the air being blown across a wet surface. In order to increase the thirst of the air, temperature must be increased or humidity decreased. Regardless of the change, if the net thirst is increased the evaporation potential is increased.

Thirst assumes that the warm, dry air is being blown directly across the wet surface. This allows for the heat energy in the air to transfer to the material (and the moisture within it), and for low humidity to be continuously re-supplied to the surface. A low vapor pressure is maintained at the surface of the material, while heat energy is applied to increase the vapor pressure of the material itself.

In concert, warmer and drier airflow promotes evaporation by influencing both material vapor pressure and air vapor pressure. This is why the best evaporation rate will occur when using the air with the greatest thirst.

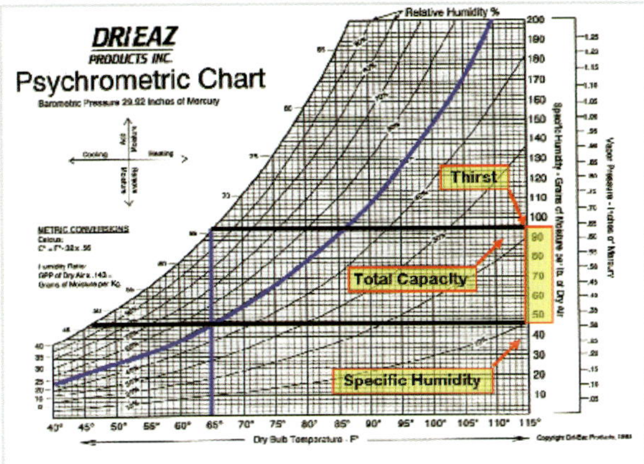

There are drawbacks to the sole use of thirst, especially if it is used to evaluate the drying environment in absence of any other psychrometric calculation. Dew point and vapor pressure must be evaluated alongside other material temperatures to ensure that there will not be a risk of condensation, or elevated humidity in other areas of the structure. In other words, the thirst of an air sample needs to be compared to the effects of cooling. If the air employed has the chance of cooling next to unaffected surfaces, the result will be an elevated humidity. Technicians must evaluate this humidity and consider its potential for causing secondary damage.

PERMEANCE

The ability for water vapor to diffuse through a material is expressed as the material's permeance or permeability. Permeability factors have been established for a variety of common building materials through testing, using the following parameters:

IMPERIAL MEASUREMENT	
Material Sample:	I sq. foot
Time:	I hour
Vapor Pressure:	Difference of I" Mercury(hg)
Unit of Measure:	Grains (I/7,000th lb)

Using these criteria, various materials have been tested and the grains of water vapor that diffuse through the material have been recorded. Each grain of water vapor that passes through the material is said to be a "perm." Therefore, materials with higher "perms" have greater permeance (permeability). The table below outlines a list of common wall assemblies and their permeance.

MATERIAL	PERMEANCE RATING
Knockdown texture & flat paint	21.53
2 coats of flat latex paint	19.54
Knockdown texture & semi gloss paint	13.33
4 mil vinyl covering (wallpaper)	00.42

Materials with permeance ratings of less than 1.0 are referred to as *vapor retarding materials* (sometimes referred to as vapor barriers). A common vapor retarding material found in homes is vinyl wall covering. Other vapor retarding materials include vinyl sheet goods, polyethylene vapor barriers, some paints and sealers and foil backing on insulation and wall paper.

Vapor retarding materials impede the drying process. Water trapped beneath and behind these materials can cause secondary damage if it is not allowed to escape in a timely fashion. Therefore, the removal of the vapor retardant material is necessary if progress is not evident.

VAPOR DIFFUSION

Vapor diffusion itself is not calculated using a psychrometric chart. However, the potential for vapor diffusion is monitored using conditions gleaned from the chart. Vapor diffusion is *the result of differences in vapor pressure on opposing sides of a material.* When vapor pressures are different, water vapor will seek equilibrium. The permeability of the material that separates the two air masses will influence the rate at which water vapor will equalize through it. The water traveling through the material will often diffuse through the material – even if cracks and crevices are not present.

Understanding the potential for vapor diffusion is particularly important when there are dramatic differences in vapor pressure. This occurs in both the initial and the final stages of restoration. Initially, vapor pressures can be very high in the affected area relative to the outdoor conditions. This will create a potential for water vapor to diffuse through external wall systems and eventually to the outdoor air. If the outdoor air is cool, this may result in elevated humidity and even condensation in exterior wall cavities.

When drawing near the final stages of restorative drying, the opposite can occur. Typically the indoor conditions are much drier than outdoor conditions, particularly in warm, humid geographical locations. This can result in water vapor diffusing through exterior walls toward the indoor environment. If the indoor air is cooler than outdoor air, this leads to elevated humidity and potential condensation.

Another reason to be aware of **vapor diffusion**: *water removal by mechanical dehumidifier may be influenced by outdoor humidity.* When the indoor air is much drier than outdoor air, water vapor diffuses through the structural components and eventually to the indoor air mass. This water vapor will then be processed by mechanical dehumidifiers indoors.

GRAIN DEPRESSION

Restorers must monitor their drying equipment to ensure that expected results are being delivered. Dehumidifiers are monitored using a calculation called grain (gram) depression. **Grain (gram)**

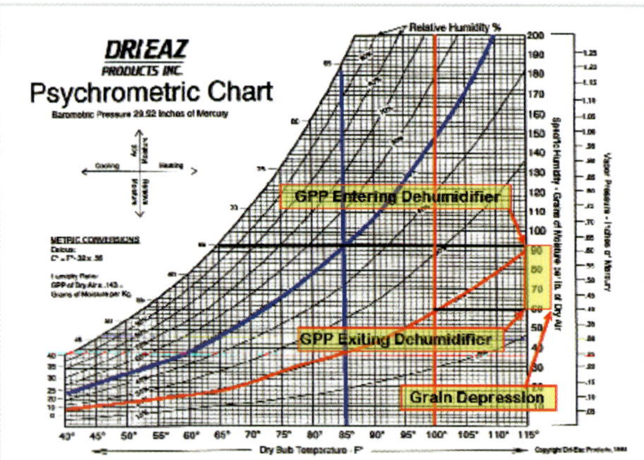

Calculating grain depression on a psychrometric chart

depression is calculated using two separate readings and a psychrometric chart.

1. Temperature and relative humidity of air entering a dehumidification system
2. Same readings of the air that has been processed by the dehumidifier (output)

Once obtained, the technician then calculates grain (gram) depression in the following way:

1. Calculation of the GPP (g/kg) for each
2. Subtract GPP entering dehumidifier from the GPP exiting.
3. This difference is the **grain depression**

Note:

It is important to consider the accuracy of the thermo-hygrometer when evaluating grain (gram) depression. Technicians must consult the product literature for their meter and consider the published performance specifications. Failure to do so will result in inaccurate conclusions.

The term grain depression is commonly used because *grains per pound of dry air (GPP), or grams per kilogram (g/kg)* is the common unit of measure in the restorative drying industry for air moisture content (**humidity ratio**). This calculation is used to evaluate dehumidifier performance, and the performance is then recorded over time to track changes in the dehumidifier's water removal. When used alone, however, grain depression does not accurately reflect the unit's performance compared to other dehumidification systems. This is true because the amount of air processed by each dehumidifier (typically **CFM**, or cubic feet per minute, or CMH, cubic meters per hour) will vary.

The definition of GPP (g/kg) – *weight of water vapor in a given volume of air* – sheds light on the fact that grain depression is simply *water removal in a given volume.* It is *not* water removal in a given period of time. For example:

- If two dehumidifiers are operating in the same room and each are delivering a 20 grain depression (2.86 g/kg), this does not necessarily mean that the two units are removing the same amount of water.
- If one unit is processing air at 100 CFM (170 CMH) and the other is processing air at 200 CFM (340 CMH), the rate at which they are removing 20 grains will be different. The unit operating at 200 CFM is removing 20 grains twice as often as the unit operating at 100 CFM.

The influence that CFM (CMH) has on actual water removal is important. It is best understood when the grain depression and CFM are used to estimate water removal in a more common measurement, such as pints of water in a 24-hour period. When these measurements are converted to pints (liters) per day, the result can be compared to the unit's rated performance. The water removal can also be compared to other dehumidifiers in the drying environment to establish performance differences.

> The formula for converting grain depression to pints of water removal per day is:
>
> CFM x 60 (minutes in an hour) = CFH
> CFH x 24 (hours in a day) = CFD
> CFD ÷ 14 (cubic feet per pound of dry air) = lbs/air
> lbs/air x grain depression = grains/day
> grains/day ÷ 7,000 (grains per pound) = lbs/water/day
> lbs/water/day ÷ 8.4 (lbs water per gallon) = gallons
> gallons/day x 8 (pints per gallon) = pints

This formula can be dramatically simplified. Because several of the factors in the formula are constant, they can be consolidated. The result is the following abbreviated formula:

> **P = CFM x GD ÷ 71**
>
> Where:
> P is Pints of Water Removal
> CFM is Cubic Feet per Minute, dehumidifier
> GD is Grain Depression
> 71 represents the abbreviation of all other constants in the original formula

This information can also be obtained using the Water Removal Estimator in the *Appendix.* The Restoration Project Manager document discussed in *Chapter 4 Inspection, Monitoring and Documentation* and included on the CD-ROM can also be used to calculate these readings.

Chapter 8

Drying Equipment

IN THIS SECTION

INTRODUCTION

The essential tools of restorative drying are air movers and dehumidifiers. Every drying technician must clearly understand the specifications and capabilities of each piece of equipment – and all safety concerns must be understood by both technicians and customers. While an untrained technician can easily plug in a piece of equipment and turn it on, a *properly trained* technician masterfully uses each tool to dry better and faster.

SAFETY

Safety is a primary concern on all restorative drying jobs. Equipment left at the site may produce safety concerns, and company liability must be limited as much as possible. Equipment is made safer by following some simple steps:

Inlet safety screen

- Observe the National Electrical Code (NEC) or other codes relevant to specific areas regarding amp draw and circuit capacity.
- Check power cords to ensure they have a grounded (earthed) electrical plug and are free of cuts and frays.
- Ensure that safety screens are in place and fully functional and free from blockage and restriction.
- Always plug equipment into Ground Fault Circuit Interrupters (GFCI) (or residual current device).
- Verify that equipment motors are thermally protected.
- Use only equipment that is UL (CE, CSA, C-UL) Listed.

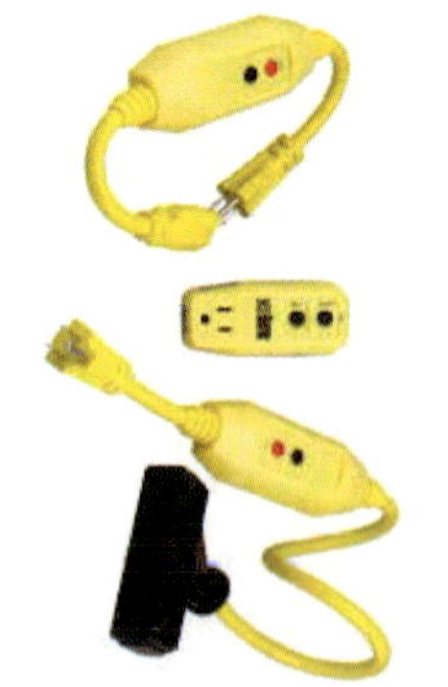

GFCIs are available in many forms.

Drawing too much power through a circuit or a lightweight extension cord is a serious safety hazard. The

National Electric Code (NEC) states that no more than 80% of the amperage available on a circuit can ever be used continuously. Most U.S. homes and commercial structures are built with several 15 amp and/or 20 amp circuits. When the NEC is applied, these circuits are used at no more than 12 and 16 amps, respectively.

NOTE

NEC EXAMPLE: when running 4 amp air movers in a structure with 15 amp circuits, no more than three units can be plugged into each circuit at a time.

Power cords with damaged or missing ground plugs cause a serious safety hazard – they must be immediately removed from service and properly repaired. On UL listed equipment, the power cord itself must be replaced if it is damaged; merely replacing the plug or a section of the cord will result in the loss of UL approval for the entire unit. Original equipment cords are purchased from the original manufacturer or official dealer of the equipment. Some manufacturers will require that an authorized service center replace the power cord, especially with larger equipment.

Additionally, inlets on air movers and dehumidifiers must not be blocked by furnishings, because this reduces performance and could cause overheating. Outlets on dehumidifiers must not be directed toward furnishings or structure items, as overdrying may occur.

Ground fault circuit interrupters (GFCI) decrease the hazard of electrical shock, among other hazards. They are commonly available in residential structures as part of the kitchen and bathroom electrical circuit. Outlets with a GFCI included contain reset and test buttons, usually in the middle of the outlet plate. When these are available, they are used for an additional layer of protection. When these are not available in the structure, restorers must use supplemental GFCI adapters. These are found as single adapters, power cords and also on temporary power distribution panels.

All equipment used in the restorative drying project must be UL (CE, CSA, C-UL) listed and approved. This ensures that the equipment has been thoroughly tested for sustained use in the indoor environment and adds a substantial amount of liability protection. Restorers must maintain their UL (CE, CSA, C-UL) approved equipment properly. This includes replacing cords that are damaged and replacing damaged inlet and outlet screens.

AIR MOVERS

Of all the tools used in structural drying, **air movers** are the greatest in number. For many years, most air movers were similar in design: the common squirrel cage blower, or carpet dryer, was used for almost every drying situation. In recent years, there have been developments in technology and design that have expanded the restorer's arsenal in air movement. Regardless of design, the role of the air mover is quite simple: *it is the tool designed to "blend" the air at the surface of wet materials with other air in the affected structure.*

By doing so, the air mover helps speed the rate at which warm air heats cool surfaces and wet, saturated air meets the intake of a mechanical dehumidifier.

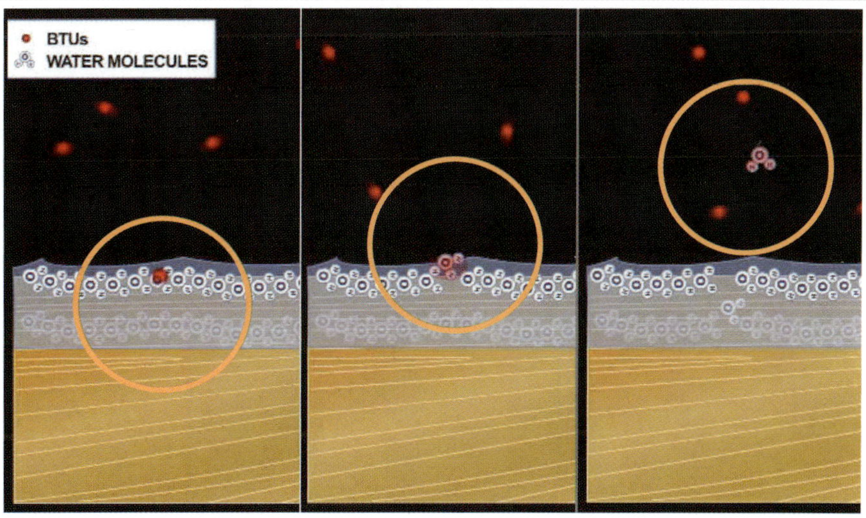

1 Energy in air and materials equalizes with water molecules

2 Water takes on energy, increasing molecule motion

3 Eventually molecules break free of each other and of the material (evaporating into the air)

In this process, air movers facilitate evaporation in two ways in a wet structure:

First, air movers **remove the** boundary layer of wet air from wet surfaces. The boundary layer is a region of high humidity near a wet material which hinders evaporation by acting as a vapor barrier. Although the ambient air in the structure may be relatively dry, the layer of air next to a wet surface can be quite humid – potentially even at saturation. Air movers remove the boundary layer by rapidly supplying drier air directly to the wet surface and thereby increasing evaporation.

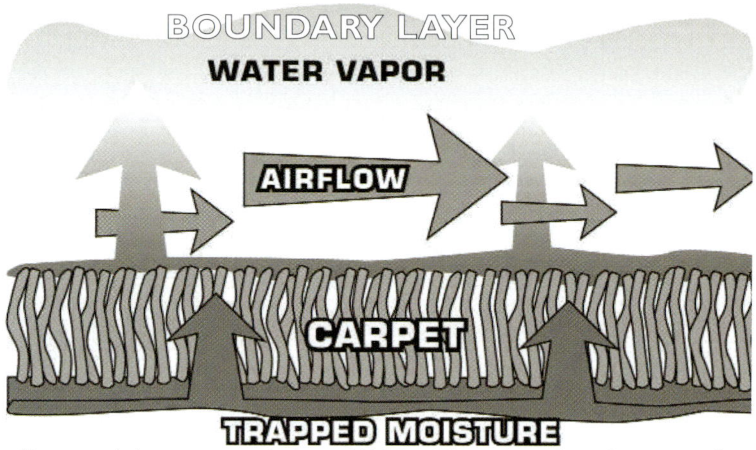

Air movers help sweep away the layer of high humidity at the surface of wet materials.

Second, air movers **circulate dry air** throughout the structure. Many structures that have suffered a water intrusion contain more than one room or area that has been affected. When this is true, it is often impractical to install dehumidifiers in each and every affected room. Air movers are used to circulate air from one affected area to the next. This eliminates the need for several smaller dehumidifiers spread throughout the structure and makes it possible to use fewer larger, more efficient dehumidifiers.

Determining How Many Air Movers to Place

Air movers are placed in the environment to ensure rapid evaporation across all affected surfaces.

Air movers are used to circulate air in the affected area. Here, two air movers are used to move air across affected materials and create exchange between two affected rooms.

Therefore, enough air movers must be installed to create airflow across every affected surface. The number of air movers necessary depends upon the number of wet surfaces, the amount of water present, and the ability for air to reach each wet material (e.g., wall cavities, behind cabinets, under contents).

Industry groups have agreed upon established numbers of air movers needed initially in an average structure. These established numbers are based upon the materials being dried, and are best understood using classes of water loss. Classes of water refer to the *potential evaporation rate from a given mix of affected structural components.* (See *Chapter 9 Drying Systems* for more details.)

A Class 1 water intrusion involves wet plywood, concrete, framing members such as dimensional lumber, and other dense materials. Very little wet carpet and underlay (pad, cushion) are present in a Class 1 scenario.

A Class 2 water intrusion typically involves whole rooms of wet carpet and underlay (pad, cushion), and wet gypsum wall board with sustained wicking of less than 24" (61 cm).

A Class 3 water intrusion has resulted in whole rooms of wet carpet and underlay (pad, cusion) and wet gypsum wall board with wicking in excess of 24" (61 cm). Generally, a Class 3 job site is a situation where water has come from overhead.

A **Class 4** intrusion results in a specialty drying situation where the affected materials contain deep pockets of saturation. The wet surface may be inaccessible to the standard air mover, water is typically bound, and evaporation will be slow.

Once the class of water loss is defined, the amount of air movement necessary is estimated using the factors below:

- In a Class 1 water loss, one air mover is needed for every 150-300 sq. ft. (14-28 sq. m.).
- In a Class 2 or Class 3 water loss, one air mover is needed for every 50-60 sq. ft. (4.6-5.6 sq. m.).
- Regardless of class, an air mover is needed for every affected area – including every offset, closet, bay window and room.

> To estimate the number of air movers needed on a job, follow these steps:
>
> 1. Determine the number of square feet (square meters) affected.
> 2. Determine the class of loss.
> 3. Divide the sq. ft. by the factors as follows:
>   - Class 1: divide sq. ft. by 150, then divide sq. ft. by 300 (will give a range necessary)
>   - Class 2 or Class 3: divide sq. ft. by 50, then divide sq. ft. by 60
>
>   In Metric, divide the sq.m. by the factors as follows:
>   - Class 1: divide the sq.m. by 14, then divide sq.m by 28
>   - Class 2 or Class 3: divide the sq.m. by 4.6, then divide sq.m by 5.6
> 4. The resulting number is the minimum range of air movers needed.
> 5. Additional airflow is necessary for building offsets such as closets and bay windows.
> 6. Specialty air movers may be necessary if sub-surfaces require airflow.

Once the number of air movers to be installed has been determined, several factors will influence their actual installation. These factors include: *the type of material affected, the degree of saturation, the accessibility of the actual wet surface, power availability* and *equipment availability.* Below are some general guidelines for installation of air movers:

- In a Class 2 or Class 3 water loss, air movers are placed every 10'-16' (3-4 m) along the wall.
- Air movers are directed toward the wall at a 15° to 45° angle, depending upon the type of air mover.
- The air mover's snout will almost touch the wall, which means within 1" (2.5 cm) of the wall but not rubbing it.
- All air movers in each area will face the same direction, whether clockwise or counter-clockwise, to ensure that air movers are not pushing against each other.

Industry groups have not determined the number of air movers needed for a Class 4 loss. This is due in part to the fact that specialized air movers are used for Class 4 water intrusions. Standard air movers generally will not create air movement directly across the wet surface in a Class 4 scenario. In addition, the affected material is dense, and will contain primarily bound moisture. Class 4 intrusions involve other difficult situations as well, including exotic materials and special or unforeseen situations.

When determining the air movement needed for a special situation, the purpose of air movement is considered: *removing the boundary layer and distributing dry air in the structure.* Enough air movement is installed to achieve these two objectives.

Air Mover Performance and Types

There are several different types of air movers available to the restorer today. Each type has been designed to meet the needs of one or more application restorers commonly encounter. Understanding which air mover is best for each application requires knowledge of each air mover's unique performance.

Air mover performance is measured in many ways, but the three main performance specifications to check are maximum **tested airflow**, **static pressure** and amp draw. Each factor serves a significant purpose or has a significant impact on the drying environment:

- **Maximum tested airflow** – the cubic feet per minute (CFM) (cubic meters per hour or CMH) of air produced by the air mover – is measured with an anemometer in a test chamber. This specification is most directly related to the air mover's intended purpose. If all else is equal,

Drying Equipment

the greater the CFM, the more influence the air mover will have in the environment. It is critical that the number considered by the restorer is the *tested* or actual airflow, and not *theoretical* or *rated*. Theoretical and rated airflow are hypothetical and will lead to inaccurate decisions when evaluating performance.

- **Static pressure** is the force of the airflow to overcome resistance. Static pressure relays how well the air mover will support air movement when there are restrictions present expressed in inches (Pascals). Air movement is restricted for a number of reasons, especially when drying a wet surface beneath and behind finishing materials. Where the application presents greater restrictions to air movement, static pressure must be higher in order to achieve appropriate flow across the affected surfaces.
- **Amp draw** is the maximum amperage used by the air mover when on high speed. In other words, this specification denotes the amount of power consumption. High power consumption can lead to limited equipment use on little available power. And, as more power is consumed, more heat will be generated by the equipment.

Every restorer knows that different drying situations require different amounts of airflow and pressure, power at the job site can often be limited, and temperature cannot be allowed to increase uncontrolled. Because of the variety of situations encountered in restorative drying, a range of **air mover types** exist to suit these individual needs. Each provides a different combination of airflow (CFM), pressure (inches of static pressure) and power consumption (amperage).

- Carpet dryers are the most common type of air mover. Carpet dryers – commonly called blowers – use a centrifugal blower to move 1,000-1,200 CFM (1,700-2,040 CMH) of air at static pressures between 1"-2" (2.5-5 cm), depending on the design. Carpet dryers (115v) typically draw 4-7 amps, depending on the model and design. Accessories for carpet dryers allow air to be ducted into small spaces, such as wall cavities and under cabinets. When attachments are used, it is generally suggested that a higher power air mover be considered (see HP Carpet Dryers below).

Carpet Dryer

- HP Carpet Dryers are specialized carpet dryers, fitted with a higher horsepower motor. Commonly referred to as HPs, they deliver a

HP Carpet Dryer

higher CFM and higher static pressure than the standard air mover. The cost, however, is a higher amp draw. Specifications for the HP range from 7-11 amps, 1,500-2,000 CFM (2,550-3,400 CMH) and 2"-3" of static pressure. They are best used as power houses for attachments such as TurboVents® or other vent/manifold systems. With these attachments, they can be used as inter-air drying systems and ventilators.

Low Presure Axial Fan

- **Low pressure axial fans** are specialized air movers used to move large volumes of air with low amp draw. With 2,000 or more CFM (3,400 CMH) drawing just 1.5 to 2.5 amps, these air movers save power while moving more air than a carpet dryer. Low pressure axial fans are efficient for drying long surfaces, open areas and carpet. Because of their low pressure, they are not as useful for pushing air into cavities and through ductwork. Because of their extremely low amp draw, these units provide the maximum amount of airflow per 15 amp circuit. Eight 1.5 amp air movers can be installed on each 15 amp circuit, delivering 16,000 CFM. Compared to the carpet dryer's 3,100 CFM per 15-amp circuit (allowing only three units per ciruit), low pressure axial fans can be advantageous, especially when drying with limited power.

- **High pressure axial fans** are used to move large volumes of air through ductwork. Commonly called ventilation fans, these air movers have the high pressure to move large volumes of air through 25'-50' (7.6-15 m) of ductwork and can generate strong negative or positive pressure. They typically use 7-10 amps (115v) while moving 2,000+ CFM and supplying approximately 3" of static pressure.

Inter-Air Drying System

- **Inter-air drying systems** are used when more pressure is needed, but air volume is less important. They are used to dry cavities such as under cabinets, wall cavities and under hardwood floors. Inter-air drying systems move approximately 100 CFM (170 CMH) and produce up to 60" of static pressure at 3 to 6 amps (115v). Since cavities have a small volume, the low CFM of the unit is effective in drying. The pressure is the main focus of inter-air dryers. A great deal of pressure is needed to push the air through lengths of tubing, through wall or other structural cavities, or to pull air up out of a hardwood floor through the cracks and crevices between boards.

High Pressure Axial Fan

High pressure axial fans are designed to facilitate ventilation.

Other specialty air movers offer varying combinations of CFM, static pressure and power consumption. There are a number of models available, and there is no doubt that additional units will be designed in the future. Technicians will most effectively match the appropriate air mover to the drying challenge when CFM, inches of static pressure and power consumption are each considered.

Air Mover Applications

Although air movers are quite simple machines to operate, there are a number of ways to apply them that will increase their impact on evaporation. It is important to consider the true purpose of an air mover: *move warm, dry air immediately across a wet surface to add energy to liquid and absorbed water and remove water vapor.* The following are several steps the restorer can take to preserve and protect the environment while increasing air mover effectiveness:

Dehumidifier exhaust ducting is attached to a clamp-on accessory rather than ducting it directly into the airmover intake.

- When drying wall cavities using high pressure air movers or inter-air drying systems, if additional airflow is required, increase the number of holes or injectors per cavity. Make these holes below the baseboard line, if possible, to reduce repairs. Use additional holes for inserting more air or to allow air to escape the cavity and reduce back pressure.
- To speed evaporation inside a wall cavity, blow dehumidified air into it through the air mover. The dehumidifier exhaust CFM and air mover CFM will be different. Therefore, do not connect the outlet and inlet of the two machines directly with ducting. Either duct the dehumidifier to the inlet of the air mover (stopping the ducting at the air mover inlet) or simply place the dehumidifier output near the air capture zone of the air mover.
- If inter-air drying systems are used on negative pressure and there is concern of contamination inside the cavity, ensure that the air output of the air mover is **HEPA** filtered. To do so, duct the output of the inter-air dryer into a HEPA filtration device. Be sure that the HEPA airflow and capture zone are big enough to handle the output of the inter-air dryer.
- Keep air movers dust free by cleaning with high pressure airflow. This reduces the spreading of dust from one site to another.
- If the air mover is severely contaminated, follow manufacturer's directions to decontaminate the equipment properly. Decontamination may not be possible by manufacturer's recommendation for some types of contaminant. When drying in highly contaminated areas is necessary, it is always best to keep air movers from becoming contaminated by first cleaning the affected area, then using air movement to promote drying.
- Use high pressure axial fans in tandem to control air pressure. For example, if negative pressure is needed in an area but the duct length is long, simply place a second air mover in line to continue moving air over long distances. This can be particularly useful when ventilating spaces in large structures and large basements with no immediate outdoor access.
- If air movers or other equipment are used in a crawlspace, the overall pressure in the crawlspace should be neutral or negative, to prevent cross-contamination into the living areas. Positive pressures can force air from the crawlspace to enter the living space.

A variety of dehumidifier sizes and types are available to restorers today.

DEHUMIDIFIERS

Dehumidification reduces the moisture content of the air. Dehumidifiers balance the drying system by removing the moisture that air movers sweep away from wet materials. This balanced drying system is best achieved when *the rate of dehumidification exceeds the rate of evaporation.*

Restoration technicians must understand how many dehumidifiers to place and which type of dehumidifier is best suited to the situation at hand in order to keep the drying system balanced. If the rate of evaporation is allowed to exceed the rated of dehumidification, secondary damage will occur. This secondary damage is discussed in detail in *Chapter 6 Materials.*

Restorers ensure that the proper amount of dehumidification has been installed by:

1) using the dehumidifier formula outlined in this chapter for initial installation, and by
2) monitoring the indoor air as drying continues to ensure that humidity is decreasing, and not increasing over time.

If humidity is decreasing, the rate of dehumidification is greater than the rate of evaporation. (Measuring humidity is discussed in more detail in *Chapter 7 Drying Science.*)

Determining How Many Dehumidifiers to Place

The science of drying has advanced to a point where a restorer can apply a few simple factors and know that a balanced drying system has been initially installed in a wet environment. The following steps are followed to determine initial dehumidification requirements, and to ensure that the indoor humidity does not linger above 60% RH (see *Chapter 7 Drying Science)* for any length of time, and is brought to 40% RH or lower within 24 hours.

Jobs vary and hidden conditions may present unforeseen or unexpected results once drying begins.

STEPS:

1 Determine the cubic feet (meters) affected or volume of air to be dehumidified by multiplying the area's length x width x height.

2 Note the dehumidifier rating (Association of Home Appliance Manufacturers, or AHAM, rating) of the refrigerant/LGR dehumidifier being used, or the process-out CFM of the desiccant dehumidifier being used. Dehumidifiers' AHAM ratings are measured in pints of water removed at 80°F (27°C), 60% RH for 24 hours.

3 Determine the classification(s) of water loss, or the initial rate of evaporation based upon the type of materials affected and degree of water intrusion. Classes were introduced earlier in this chapter, and range from Class 1 to Class 4. Classes are discussed in more detail in *Chapter 9 Drying Systems.*

4 Check the dehumidifier factor table, page 100. These factors are based on how many cubic feet of air will be dehumidified by one pint of dehumidifier capacity, or, in the case of desiccants, the number of air exchanges per hour needed.

Once all the factors are known, restoration contractors then determine the initial amount of dehumidification needed:

1 Use (page 100) to determine the "factor" that will be used. The class of water loss and type of dehumidifier are both noted in the column and row headers of the table.

2 Once the factor has been identified, use the following formula for LGRs and conventional refrigerants:

Imperial: cubic ft. ÷ factor = AHAM pints required

Metric: cubic m. ÷ factor = AHAM liters required

3 For desiccant dehumidifiers, use the formula:

Imperial: factor x cubic ft. ÷ 60 = CFM required

Metric: factor x cubic m. = CMH required

4 The product of the formula represents the minimum CFM (desiccants) or AHAM pints (refrigerants) required. This total is then compared to the dehumidifier's published rating. The appropriate number of dehumidifiers are then installed to ensure that the minimum requirement is met. For example: if 330 AHAM pints are required, and 100 pint dehumidifiers are available, four are installed.

DEHUMIDIFIER FACTORS TABLE

	Class 1	Class 2	Class 3	Class 4	Units
Conventional refrigerant	100	40	30	N/A	cu. ft. per pint
	6	2.4	1.8	N/A	cu. m. per liter
LGR	100	50	40	50	cu. ft. per pint
	6	3	2.4	3	cu. m. per liter
Desiccant	1	2	3	2	AEH
	1	2	3	2	AEH

The table above is used to determine the "Factor" used for dehumidifier formulas (imperial measures in black, metric in color).

Therefore, it is necessary to use the following guidelines when applying the initial dehumidification formula:

- Formulas are intended for initial installation only. After the initial installation, the dehumidification may need to be increased or decreased based on change or lack of change to psychrometric and other moisture readings.
- If LGRs and conventional refrigerants are mixed on a job, use the formula and steps for conventional refrigerants.
- If the formula calls for a certain CFM or AHAM pint (liter) removal, add enough dehumidification to the site to cover that capacity. Providing less dehumidification than needed can cause an unbalanced drying system and secondary damage.
- The required CFM or AHAM pint removal should not be doubled. Placing too much dehumidification on a loss creates heat problems.

The DrizAir 1200 is an example of a conventional refrigerant.

The factors from the dehumidification chart were determined by industry experts and have been accepted by the IICRC, most restorative drying companies and several insurance companies. The dehumidification chart was designed to create a balanced drying system that achieves the goal of 40% relative humidity or less within 24 hours. A relative humidity less than 40% is generally a good drying condition. The LGR and conventional refrigerant factors themselves are the number of cubic feet of affected air each pint of dehumidification will handle.

The desiccant factors are the number of air exchanges per hour (AEH) needed for each class of loss. *For example: a Class 3 loss will require three air exchanges per hour through a desiccant dehumidifier.*

Principles of Dehumidifier Operation

There are a number of technologies available for mechanical dehumidification and a number of specialized systems that provide dehumidification while never removing moisture from the affected air. Each of these systems fills a specific need for the professional restorer.

Conventional Refrigerant. Conventional refrigerant dehumidifiers are the most common type of portable dehumidifier in the restoration industry. They remove water from the air through the process of condensation. Because of this fact, they are more efficient in higher humidity conditions. Refrigerant dehumidifiers work best between 70° and 90°F (21°C-32°C). Conventional refrigerant dehumidifiers contain a few key components:

- The **cold coil** (also known as the evaporator coil) is where the water from the air condenses. The cold coil is typically 20° to 30°F below the dew point temperature of the ambient air. Cold coils cannot get colder than 28°F (-2°C).
- The **hot coil** (also known as the condenser coil) is where the dehumidifier purges waste heat, which makes the air exiting the dehumidifier thirstier.
- The compressor is used to pressurize the refrigerant into the hot coil and pull refrigerant from

the cold coil. Of all parts of a conventional refrigerant, the compressor consumes the most electricity.

- The capillary tube connects the cold and hot coil and maintains the pressure difference between the two coils.
- The blower pulls air across the cold and hot coils.
- Electronics control defrost sensors, timers and other options.
- The defrost mechanism melts ice off the coils when the cold coil gets below 32°F (0°C).

It is important to note that there are two main types of defrost control systems:

- Time temperature defrost systems turn off the compressor and keep the blower running, pulling air across the frost and melting it over time. Dehumidifiers with time temperature defrost system will work down to 55°F (13°C). At colder temperatures, these defrost systems are not able to defrost effectively.

- Hot gas bypass defrosts use the refrigerant in the "hot coil" to quickly melt the ice, dumping it into the cold coil through a bypass valve. Dehumidifiers equipped with hot gas bypass defrost systems can operate in very cold conditions – as low as 33°F (.6°C). Another benefit of hot gas bypass systems is that they defrost quickly. Instead of waiting for the ambient air temperature to melt the ice from the coil, the hot gas melts the ice rapidly. The unit will spend less time in defrost mode and therefore will operate more effectively.

Conventional refrigerant dehumidifiers work by pulling air across the cold coil. As air moves across the cold coil, its temperature drops below the dew point. When the air is cooled below dew point, water is deposited on the coil. The wetter the air is, the easier it is for the air to release moisture. The drier the air is, the harder the dehumidifier must work to remove moisture. If the air becomes too dry or too hot, the dehumidifier can no longer get the air below dew point temperature effectively, and performance decreases.

One of the main goals during structural drying is to reduce the relative humidity to below 40% in the affected area. Clearly, the restorer must create drier air within the structure in order to meet this goal – but creating dry air reduces the dew point temperature of that air. This affects the use of refrigerant dehumidifiers in two ways:

Delivering dehumidified air directly to an affected structure cavity using lay flat ducting.

- Lower dew point temperatures mean there is less water in the air, and less water in the air means less water removal by the dehumidifier. The drier the air, the less water the dehumidifier will remove.
- Lower dew point temperatures also mean there is less energy in the air. Suspended water molecules carry a relatively high amount of energy. When air is drier, this energy is not present. Therefore, the cold coil is subjected to less heat when air is drier, and it begins to run much cooler. This leads to more frequent defrost cycles, and it highlights the need for efficient defrost systems (such as the hot gas bypass).

In order to improve performance of refrigerant dehumidifiers as the air becomes drier, the restorer must provide the dehumidifier with cooler air. Most refrigerant dehumidifier manufacturers publish an ideal temperature range of 70° to 90°F (21° to 32°C).

- The higher temperature (90°F/21°C) is optimal when humidity is high.
- The lower temperature (70°F/32°C) is optimal when humidity is low.

As humidity drops, refrigerant dehumidifiers will work most effectively if the temperature is also slowly dropped from 90°F (32°C) to 70°F (21°C).

Because of this principle, a refrigerant dehumidifier's performance can be boosted by feeding the intake from an air conditioner supply vent – which provides the dehumidifier with air that is closer to dew point. This way, the dehumidifier does not have to work to cool the air as much to reach dew point, and to remove moisture.

Cooler air can help increase dehumidifier performance.

However, this setup must only be used when the air is relatively dry (below 60 GPP, 8.57 g/kg). Restorers will know that this opportunity exists when air conditioning systems are no longer removing moisture (no longer have a grain depression). If this setup is used when air conditioning systems are still removing water, then the air provided to the refrigerant will not only be cooler, it will be drier – and the drier air will limit dehumidifier performance.

Conventional refrigerant dehumidifiers under normal temperatures are effective in reducing high humidity to a normal level (approximately 40% relative humidity). They are not capable of reducing the humidity to a level appropriate for supporting evaporation when the restoration project includes large amounts of **bound water**. For this reason, conventional refrigerants are not used on Class 4 water intrusions.

Low grain refrigerant

Low grain refrigerant. While conventional refrigerants are not capable of providing significant water removal once the humidity goes below 40 GPP (5.72 g/kg), low grain refrigerant (LGR) dehumidifiers will. In fact, LGRs are much more efficient than other types of dehumidifiers in most drying situations. They work well in high humidity and work efficiently in drier conditions. LGRs will produce lower GPP humidity, and therefore lower vapor pressures, than conventional refrigerant dehumidifiers.

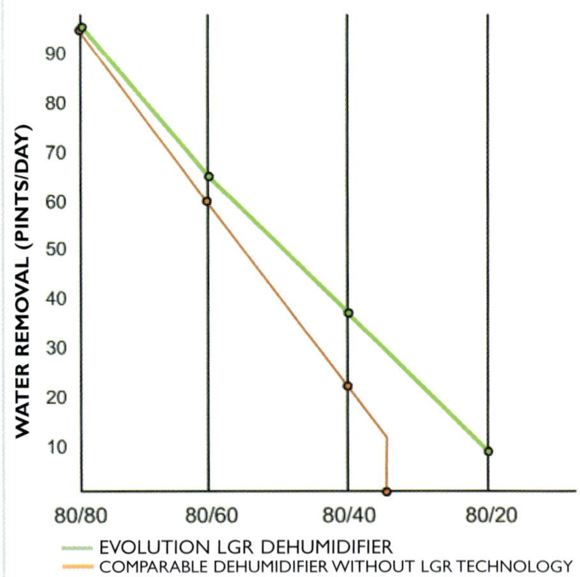

LGRs are capable of removing significatly more water in dryer conditions.

An LGR achieves this higher efficiency by incorporating a pre-cooling device, which supplies the unit with cooler air to process. The supplemental precoolers used do not require electricity, and they make it easier for the cold coil to remove water from the air. This results in a much broader operating range for temperature and humidity, and dramatically increased water removal while using less power. There are three types of pre-coolers available:

- The **heat pipe** is a looped system that surrounds the cold coil. Refrigerant within the pipe collects heat from the incoming air, and the refrigerant then heats and evaporates. The heated, evaporated refrigerant is transferred to the back side of the cold coil where it is cooled and condensed. This cooled refrigerant is then transferred to the front of the coil. This cycle repeats over and over, continually supplying cooled refrigerant to the front of the cold coil and pre-cooling the incoming air.

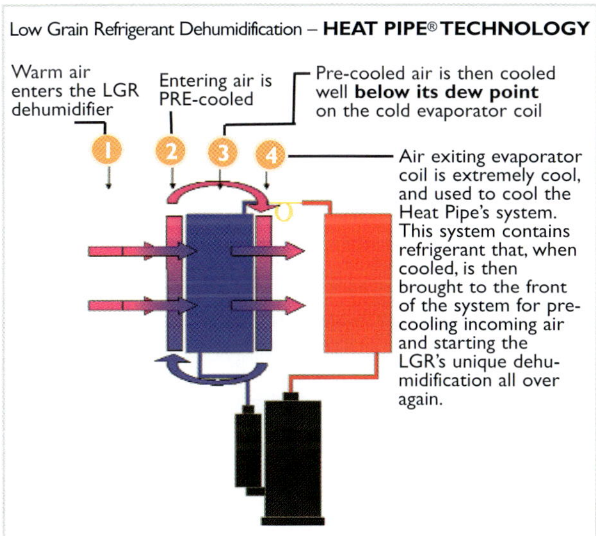

- A **thermal siphon** is a similar technology to the heat pipe, but is designed to work with a "stacked" hot and cold coil and fit into a much smaller dehumidifier chassis.

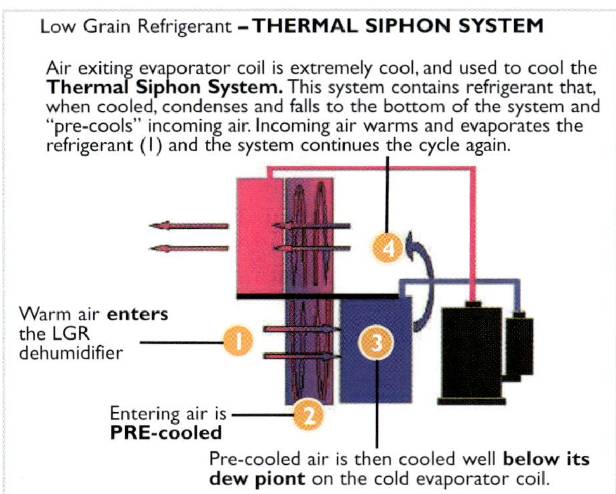

- An **air to air heat exchanger** is a plate-type air to air exchanger, made of corrugated plastic or aluminum, which places warm incoming air immediately beside the cooled air that's leaving the cold coil. This results in the removal of heat from the incoming air, which is then transferred to the air entering the hot coil.

In many ways, LGRs are similar to conventional refrigerants. For instance, LGRs also work most efficiently in conditions of 70° to 90°F (21°-32°C). Additionally, an LGR's performance can be boosted further in dry conditions by feeding the intake from an air conditioner supply vent, as long as the air conditioner is not drying the air.

In other ways, LGRs are very different than conventional refrigerant dehumidifiers. Because of the pre-cooling effect of the supplemental components, their energy efficiency, water removal potential and low humidity performance are much greater. For these reasons, LGRs are the best choice for most drying situations.

Desiccant. Desiccant dehumidifiers depend upon a chemical attraction to remove water from the air. Because they do not use condensation to remove water, they are not limited by low dew point temperatures; therefore, they produce the lowest vapor pressure of any dehumidifier (when used properly). The best conditions in which to use a portable desiccant dehumidifier are: *cool and/or dry environments.* The cooler and drier the air going into the desiccant, the drier the air will be going out. If conditions are cold, i.e., cooler than 60°F (16°C), the desiccant dehumidifier is the only dehumidifier that will efficiently remove moisture.

Desiccant dehumidifiers range in size and configuration dramatically. Desiccant dehumidifiers used in restorative drying may be as small as a roll-on suitcase or as large as two semi tractor-trailers. Smaller desiccant dehumidifiers are more common in the restoration industry today, yet larger desiccants are being used more and more frequently. Large desiccants are often self-contained dehumidification systems, utilizing onboard generators and running on propane or diesel fuel. When large catastrophic events occur, such as hurricanes, large desiccants can provide a means of dehumidification when little or no power is available.

Regardless of their size, all desiccant dehumidifiers contain the following key components:

- A **desiccant wheel** with a grid of small air passages, similar to a large bundle of coffee straws, is impregnated with silica gel (the same silica that is used in the packets that keep shoes dry). The desiccant wheel rotates slowly through the different airflow zones of the desiccant, as it works to remove moisture. The size of this rotor and the speed at which it rotates varies, depending on the unit's specifications for target humidity and air volume. Most of the cost of desiccant dehumidifiers is in the desiccant wheel.

Desiccant wheel

- The **blower** on a desiccant moves the air through the desiccant wheel. Most portable desiccant dehumidifiers use a single blower. Normally, 75% of the air is dried and sent through the wheel, exiting the unit as warm, dry process air. The remaining 25% of the air is used to remove moisture from the wheel, exiting the unit as warm wet reactivation air.
- A **heater** boosts the temperature of this reactivation air, so that it can drive out the moisture that has collected on the desiccant rotor. Hot air removes moisture from the silica by adding enough energy to the water molecules to overcome the chemical attraction they have to the desiccant media (e.g., silica gel). This function of a desiccant dehumidifier requires a high level of power consumption.
- **Electronics** include switches, sensors and a small motor that rotates the desiccant wheel and controls the unit's operation.
- **Outlets and inlets** are necessary to control the desiccant's airstreams. The *process inlet* – where 75-100% of the unit's airflow enters – typically contains a filter. *Process outlets* – which are normally located on the opposite side of the machine – are where 75% of the air that entered the process inlet is returned as warm, dry, processed air. The *reactivation outlet* may be on either side of the desiccant dehumidifier; here, 25% of the process inlet air or separate intake is exhausted as extremely warm, humid air. This air is normally exhausted outside. In summary:

Process air inlet: 75-100% of the air enters this inlet. (Provide cool, dry air for best performance.)
Process air outlet: 75% of the unit's airflow. (Provides the dry air for drying.)
Reactivation air outlet: 25% of the unit's airflow. (Warm, wet air that should be ducted outside.)

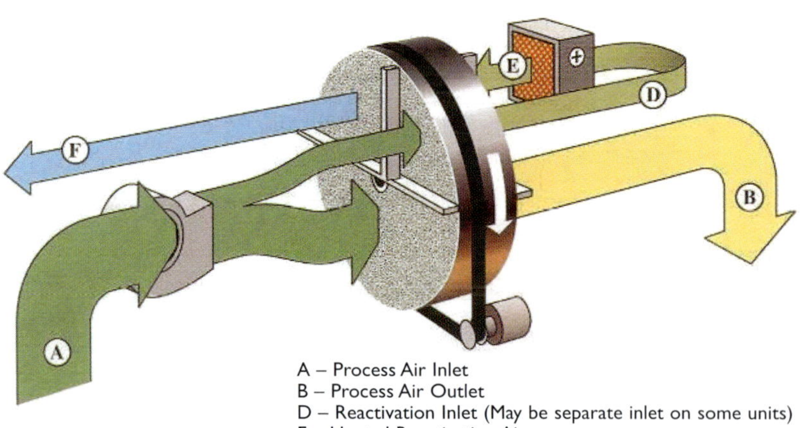

A – Process Air Inlet
B – Process Air Outlet
D – Reactivation Inlet (May be separate inlet on some units)
E – Heated Reactivation Air
F – Reactivation Air Outlet

Airflow through a small desiccant dehumidifier.

Desiccant dehumidifiers have many applications in restorative drying, but they are most commonly used to dry materials such as hardwood floors, plaster and lathe walls, or other dense materials. Desiccants effectively accelerate drying inside walls when used in conjunction with air movers. This procedure is especially effective with inter-air drying systems, which typically possess low airflow rates; feeding the inter-air system with dry, desiccant-processed air will maximize drying efficiency.

Since most desiccants do not collect water in a pan for automatic pump-out, the damp reactivation air is vented from the structure in an air stream. In cases where desiccants are used in this manner, there will be a pressure differential in the drying environment. This occurs because air is split into two air streams inside the desiccant dehumidifier, and is then delivered to two separate spaces as process air and reactivation air.

Desiccant dehumidifiers are far different from refrigerants – and in some ways can be quite versatile. There are a number of ways to increase a desiccant's efficiency by maximizing airflow and providing appropriate temperature and humidity. Some of the ways that performance can be enhanced with a desiccant are highlighted below:

- **Keep the reactivation air duct as short as possible.** Long ducting will restrict airflow through the reactivation chamber and will inhibit water removal from the rotor. Long duct runs also can cause the reactivation air to cool below dew point and can fill the duct with water. This can cause damage to the dehumidifier and the structure.
- **Duct the process air inlet from the coolest, driest air available at the job site** – whether it's outside air, cold air from the air conditioner, or air from the unaffected area. Cold air enhances desiccant performance for two reasons:
  - Cold air is normally drier air – and the drier the process air going into a desiccant, the drier the process air out
  - The rotor is dried using very high temperatures and must be cooled after reactivation, as a hot rotor will not collect water vapor. The process inlet air is used to cool the rotor - and the cooler this air, the faster the rotor will return to operable temperature.
- **In cases where the desiccant is too far from a window or exit, duct the warm, wet air into a refrigerant dehumidifier.** The refrigerant can easily collect much of the moisture from this warm, wet air.

Many technicians are concerned that if a desiccant is used, it will dry the structure too much. It *is* true that a desiccant will dry the air more than a refrigerant or an LGR, but daily monitoring prevents problems with drying. If daily monitoring is not possible, then a desiccant is used in conjunction with a humidistat, which is similar in operation to a thermostat. A **humidistat** will govern the dehumidifier's performance and can be set to prevent the unit from creating extremely low humidity. *A humidistat must be used on a desiccant dehumidifier designed for this application.* When desiccants are powered down, a delayed switch will continue to run the blower and cool the heating elements. If this is not allowed to occur, damage may result to the heater and the desiccant rotor.

Using a refrigerant dehumidifier to remove moisture from reactivation outlet

Note: Additional discussion on the use of desiccant dehumidifiers in different drying systems (eg., open drying systems, closed drying systems) can be found in *Chapter 9 Drying Systems.*

4 WAYS TO INSTALL A PORTABLE DESICCANT

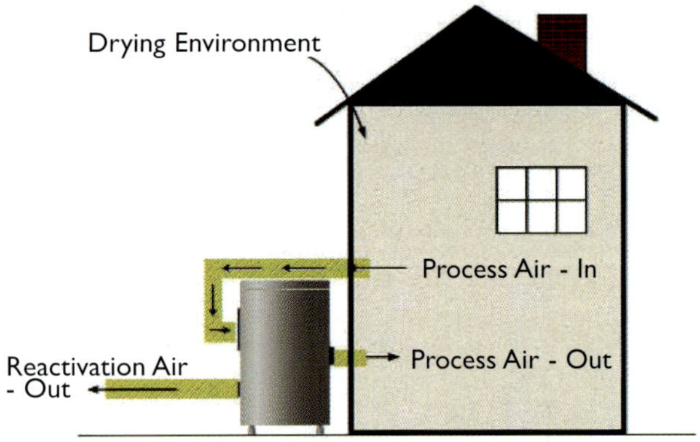

Outside Recirculation: Often used when drying environment is drier and cooler than unaffected areas. Also used when affected area is small or inaccessable.

1

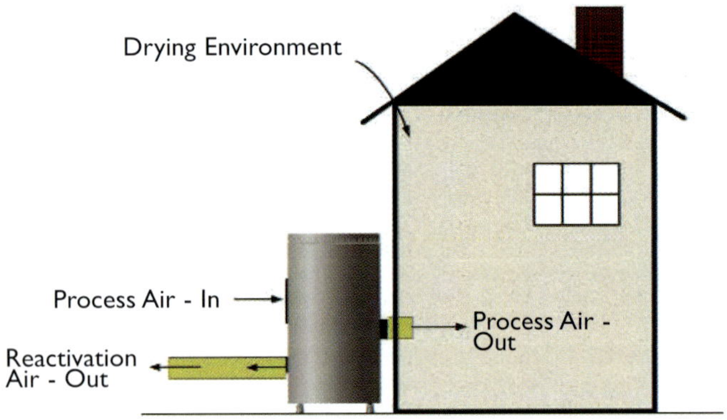

Outside Exchange: Used when unaffected air is cooler and drier. Also used when affected area is small, inaccessible, or for a small tented area such as a hardwood floor.

2

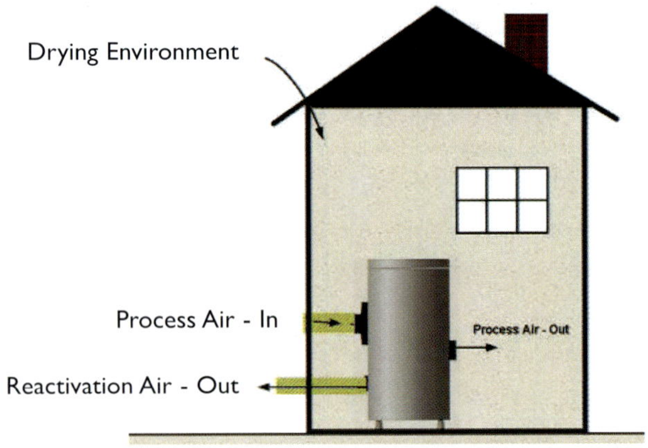

Inside Exchange: Used when outdoor air is cooler and drier. Used when unit can be installed inside the affected area.

3

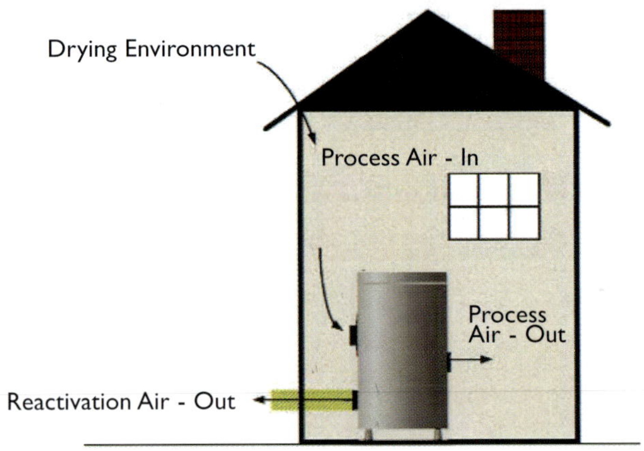

Inside Recirculation: Used when affected area is cooler and drier than unaffected. Used when unit can be installed inside affected area.

4

Heat and air exchange. In recent years, a number of specialized tools and techniques have emerged in the restorative drying industry. One technique, the heat and air exchange system, employs outdoor air and supplemental heat to facilitate evaporation and corresponding humidity control.

The process is very similar to those used by lumber mills during kiln drying. Very high temperatures are used to increase the vapor pressure associated with free and bound water in the structure (see *Chapter 7 Drying Science*). As moisture evaporates, wet air is exhausted to the outdoor environment to control indoor humidity.

Exhausted air is replaced by heated outdoor air. The heated outdoor air is relatively dry, and is promptly placed in contact with the wet surface to continue the cycle until materials reach their drying goal.

A major difference between kiln drying and structural drying is the restorer's access to the number of affected surfaces. In a kiln, lumber can be accessed on all sides. Materials are separated to ensure that the hot air can effectively heat the entire material, thereby maximizing the internal vapor pressure associated with water in the material. Air is then evacuated from all sides, ensuring that all evaporated water vapor is removed.

In a wet structure, restorers often do not have access to all sides of affected materials unless invasive methods are employed. If all wet surfaces are made immediately available to the heated air, the method can work effectively. This requires the removal of several finish materials, the application of several inter-air drying systems, or access to sub-surfaces by other means.

If any affected surface or sub-surface is not in immediate contact with the hot and dry air, water vapor can collect. Humidity can increase and may eventually result in secondary damage. If surfaces (which are in direct contact with hot, dry air) are allowed to dry at a greater rate than sub-surfaces, damage such as checking, splitting or warping will occur.

When considering the use of heat and air exchange, restorers need to take into account the temperature limits set by occupants, sensitive contents, changing weather and psychrometric factors. Heat and air exchange can be used to dry structures in certain circumstances, after each of these factors have been addressed. In cases where a heat and air exchange is determined appropriate, outside air acts as the dehumidifier. Forced air heaters and air exchange devices offer the ability to use outside air that is made more thirsty to remove moisture from the structure.

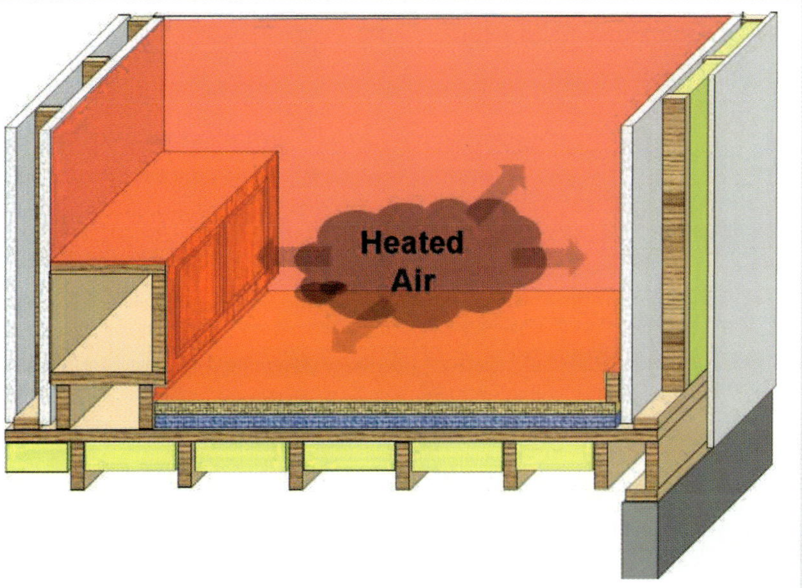

Heating the air in a structure will only affect the materials in direct contact with the heated air. Many potentially wet surfaces will not be in direct contact with the heated air, unless other measures are taken (i.e., using an inter-air drying system, opening wall cavities).

A forced air heater is installed to heat a crawlspace.

Forced air heaters used in restoration are typically indirect fired furnaces. Indirect fired furnaces are placed outside a structure and pump heat into the affected area either by blowers or liquid heat exchangers. Diesel or propane burners create the heat to increase the temperature inside the structure, but noxious fumes are vented outside.

Forced air heaters are most effective when the outside air is dry. Warm, dry air is ducted into the structure from the heater, gathers moisture from the structure, and is then vented out the opposite end of the structure with forced air.

When using forced air heaters, efficiency and safety can be maximized by:

- Ensuring that fuel is contained in safety certified containers that are stable and secured.
- Using thermostats properly to keep temperature under control.
- Controlling pressures. (Always provide a method for hot air and evaporated moisture to be removed from the structure.)
- Ensuring that the furnace flue is working properly, and that noxious fumes are not ducted inside the structure.

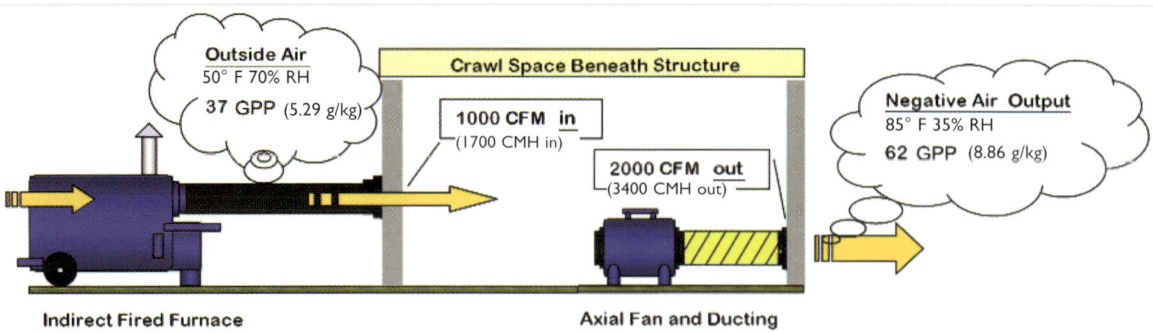

An example of installation of heat and air exchange

Note: An extensive discussion on the use of heat and air exchange (e.g., open drying system, combination drying system) can be found in *Chapter 9 Drying Systems*.

Proper Placement of Dehumidifiers

Dehumidifiers are placed in areas where the most rapid evaporation occurs; this is where they can perform most effectively. Air movers are employed to ensure a proper amount of circulation between the affected areas, and to aid evaporation. If proper circulation cannot be achieved, multiple dehumidification systems are used to ensure even humidity control. It is critical that the restorer evaluate the environment to ensure that humidity is effectively controlled in all areas of the structure. Restorers accomplish this using the following techniques:

- ***If multiple dehumidifiers are used on a job, space them evenly throughout the structure.*** Humidity control is much more effective when there are more capture zones, or if circulation is better.
- ***Air should not be blown directly into dehumidifier intakes.*** This changes the rate of airflow across the dehumidifier coils and reduces performance.
- ***Place dehumidifiers near the air stream (but not blocking the air stream), with the air blowing across the back of the dehumidifier.*** The air blowing across the side of the dehumidifier (opposite the intake) evens out the exterior temperatures of the dehumidifier and immediately disperses dehumidified air, increasing the dehumidifier effectiveness.
- ***Do not place dehumidifiers on top of hardwood (or other high value) floors.*** Dehumidifiers can leak if not properly maintained, or if components fail (e.g., water pumps).
- ***Do not blow warm, dry air at finished, hard surfaces.*** This can overdry and damage surfaces.

AIR FILTRATION DEVICES

Air filtration devices (AFDs) are needed in water restoration work whenever high levels of particulates are being added to the air. AFDs are especially important if there is strong air movement, contaminated water or a contaminated structure. Typically,

Install dehumidifiers in the path of airflow.

at least one of these conditions is true in every drying environment.

Air filtration is more important when building occupants are at a significant risk for discomfort from elevated particulate in the air. This may be the case if occupants are very young, elderly, or have respiratory problems or a compromised immune system. In these cases, air filtration devices must be employed.

AFDs are used whenever drying in a structure with a known contaminant. Special care – including proper cleaning and air filtration – is used when wall cavities and flooring surfaces or sub-surfaces are known to carry a high degree of organic debris, soils, allergens or other undesired particulate.

Blowing outlets from dehumidifiers directly at furnishing can cause damage.

AFDs – often called negative air machines, air scrubbers, HEPA **units** or **air filters** – offer several stages of filtration, including:

- **Pre-filters** – Designed to catch large particles, replaced daily or at least with each job. Often available in two stages for varying particle size.
- **Carbon filter** -- Often optional. Designed to absorb organic vapors, thereby reducing odor.
- **Primary filter** – Often made of HEPA (high efficiency particulate air) media. Designed to capture 99.97 % of particles down to 0.3 microns in size.

Using multiple, smaller AFDs will greatly increase air filtration effectiveness.

Air Filtration Devices

Before installing units to manage the aerosolizing of particulate, it is important to understand the way in which air filtration devices are rated. Similar to the principle of removing moisture with dehumidifiers, air filtration devices must be capable of removing particulate at a rate greater than it is being added to the air. This will result in a weighted balance in favor of the filtration device and will ensure control of airborne particles.

The predominant rating used in the restorative drying industry for air filtration is a **filter efficiency rating**. Filter media is tested to ensure that it captures a high percentage of particulate at a given size. A HEPA rating denotes that the filter removes 99.97% of particulate at 0.3 microns in size. However, an air filtration device employs a blower, a HEPA rated filter, various gaskets and seals, and a housing – and all of these components are not necessarily tested in conjunction with each other.

The efficiency of the filter is typically published as is the CFM of the blower. However, when the filter and blower are both installed, CFM will decrease and filter efficiency will also be affected. The result can be a performance much lower than expected. *For example: Filters tested to deliver HEPA at 200 CFM will not deliver the same efficiency at 1,000 CFM. Additionally, blowers rated to deliver 1,000 CFM will not do so when they are restricted by a HEPA filter.*

Therefore, when evaluating an air filtration system's potential effectiveness, the components must be tested after complete assembly. This testing should focus on the efficiency of the filter and the actual CFM delivered by the machine.

CADR (Clean Air Delivery Rate) is an AHAM testing standard, similar to those used for dehumidification water removal. CADR testing requires that the air filtration device be tested with *all* components assembled. The actual CFM and filtration efficiency are measured, and after the CFM and filtration are known, the following formula is used to establish the CADR rating:

CFM X FILTRATION EFFICIENCY = CADR

Using this formula, the amount of clean air delivered by the unit is calculated and expressed as the unit's CADR. This rating is critical to the needs of the restorer. When mitigating aerosolized particulate, both airflow and filtration are necessary.

A filter that theoretically removes 100% of particulate, but has no airflow, will actually remove no particulate. Using the CADR formula, the unit would have 0 CADR. A filter with 2,000 CFM but no filtration will remove no particulate. This unit also would have a 0 CADR. In other words, effective air filtration devices remove a high amount of particulate at a high CFM. For example, a unit that delivered 500 CFM and 99.97% efficiency would deliver a 500 CADR (500 CFM x 99.97%). Effective air filtration devices have a high CADR.

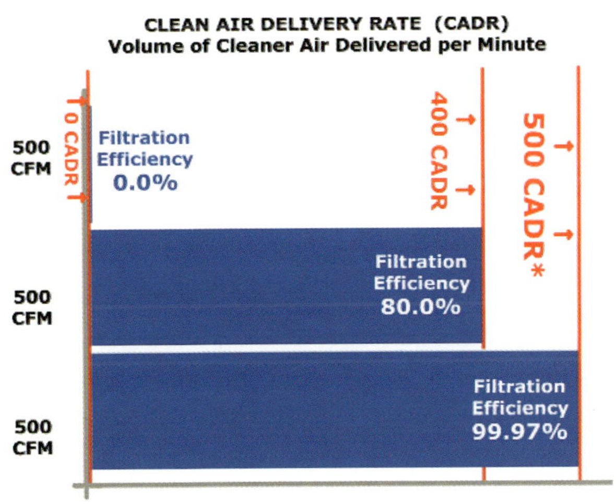

*As confirmed by third party testing. (The maximum airflow tested and rated by AHAM is 450 CFM)

Determining How Many AFDs to Install

Determining the amount of air filtration needed is very similar to how desiccant dehumidifier CFM is determined. The factors involved are:

- **Cubic feet or cubic meters** of air in the area to be filtered
- **Air exchanges necessary** (ranging from 4 to 10, depending upon conditions)

CHANGING A HEPA FILTER

1

While wearing proper PPE, open unit outdoors and remove filter. Place directly into a suitable container. Keep unit running.

2

Seal container immediately. Continue until all filters are removed. Place a second bag over the first, and seal.

3

After all filters are removed, clean interior of housing.

4

Install new filter only after cleaning unit thoroughly.

- **Math** (Multiply the cubic feet by 4 to 10, then divide by 60; this will provide the CFM/CADR needed to effectively filter air in the affected area.)

This formula can be expressed as:

Cu. Ft. x AEH / 60 = CFM/CADR Necessary

Where:

Cu. Ft. = Cubic feet of air in the affected area

AEH = Air exchanges necessary (4 to 10)

60 = Minutes in an hour, to convert air exchanges per hour to cubic feet per minute

It is important to place AFDs evenly throughout the structure in order to capture the most particles.

AFDs are most effective when installed as multiple units with lower CFMs – as opposed to fewer units with higher CFM – to greatly increase the capture zone in the affected area. If this is not possible, and only one large AFD is used, it must be placed in a central location. AFDs can also be placed in the air stream of the structure's HVAC system or in the air stream of the air movers to increase the capture zone of the AFD.

When changing the pre-filters of AFDs at the loss site, follow these steps to prevent spread of contaminants:

1. Keep the AFD running during the entire process.
2. Place the containment bag near the filter that will be changed.
3. As the filter is removed, place it in the bag and seal immediately.
4. Replace the filter with a clean one immediately, to prevent the HEPA filter from filling with large particles.

ELECTRICITY

Electricity is an absolute necessity for effective structural drying and is also a major source of potential safety hazards. Nearly all of the equipment on a restorative drying job site runs on electric power, so technicians must know basic principles of electricity. With this understanding, restorers will make educated decisions regarding the amount of power available, accessing it safely, managing heat loads, using supplemental power such as generators, and directing customers on power consumption costs.

Electricity is a life blood to the restoration effort. Without it, the tools used to facilitate evaporation and dehumidification cannot operate. Therefore, limited power directly hinders the effectiveness of a restorer. Being able to identify power sources and use them to a safe but maximum level is critical.

The safe maximum requires a great deal of respect. Federal, local and provincial laws and codes prevent the average person from meddling with power sources, and many of the procedures necessary for maximizing power availability require a licensed electrician. Technicians must *never* try to modify *any* electrical circuit in a home or in a piece of equipment if they are not licensed and authorized.

Principles of Electricity

***Volts*, *Amps* and *Watts*.** To understand how electricity works, the first step is to know what electricity is:

> *Electricity is a flow of electrons inside an electrical conductor that can be used to do work.*

In restorative drying, this work is usually a motor turning a fan or compressor, or a heating coil producing heat. Electricity can be measured a number of ways, including:

- **Amperage:** The strength of an electric current measured in amperes (amps). Electrical circuits are designed and rated to carry a certain number of amps, and they should not be loaded with more than 80% of that amperage. *For example: a 15 amp circuit should not carry more than a 12 amp load.*
- **Voltage:** Electrical potential or "push" measured in volts. U.S. residential voltage is typically 110-120 volts. Combining two residential lines of the same polarity will result in a voltage of 220-240.
- **Watts:** A unit of electrical power. Watts are the product of the strength of current (amps) and the electrical potential (volts), and are commonly referred to as kilowatts (increments of 1,000 watts).

The relationship between amperage, voltage and watts can be expressed as:

amps x volts = watts

Power Availability

In the average residential structure, power is supplied in the form of 230 volts and between 100 and 200 amps. When a structure is being supplied with 230 volt service, yet the power is being consumed at 115 volts, the available amperage is doubled. In other words, 200 amp, 230 volt power is equivalent to 400 amp, 115 volt power.

Several devices are available to step down the 230 volt power. These devices are often referred to as temporary power distribution panels, or power boxes. Power boxes are normally intended for construction use and are designed to be hard wired to temporary power poles installed by local utility companies. However, power cords are available that will allow the power box to be plugged into the 230v outlets available in many modern homes (e.g., range outlet, dryer outlet).

Temporary power supply boxes must only be used when the box and the power cord have both been constructed by a licensed electrician, are UL listed and approved, and do not require the technician to make any further modifications.

Another resource for additional power is the portable generator – appropriate for the many circumstances in which power availability is either minimal or non-existent. Portable generators range in size dramatically, with smaller units capable of powering only a single 15 amp circuit, and larger units capable of powering an entire home.

Whenever a generator is used, it must be properly sized to meet the power demands of the drying equipment. To calculate the needed generator size, it is necessary to first evaluate the structure and its equipment requirements. Once the amount of equipment necessary has been established, the following formula can be used:

Killowatts Required to Run the Equipment

Kw = (Ta x V) / 1,000

Ta = total amperage requirement
V = voltage requirement
1,000 = watts per kilowatt
Kw = kilowatts per hour

Generators are typically rated in kilowatts per hour (kw). The formula above will calculate the total number of kilowatts necessary to run the required equipment. At times, it is necessary to use only as much equipment as a generator can support – and that is limited by the capacity of the generator itself. To calculate how much equipment can be run by a generator, the following formula can be used:

Generator Capacity in Amps

$$Ta = \frac{Kw \times 1{,}000}{V}$$

This formula will calculate the total amperage available, given the kilowatt rating of the generator. As with any electrical circuit, the capacity of the generator must not be used at a 100% load continuously. The National Electric Code (NEC) states that *electrical circuits must be run at no more than 80% of total capacity*, if run continuously.

Electrical Costs

Building owners will often share concerns about the cost of electricity consumption. While equipment will consume a great deal of electricity, the total cost is usually far less than most building owners would estimate. Additionally, this cost is normally a covered expense under insurance policies. Most insurance companies will reimburse the building owner based on a comparison of previous year and current year utility bills. If a more immediate answer is necessary, an estimate is calculated.

Estimated Cost of Electricity

$$\text{Cost/Day} = \frac{(Ta \times V)}{1,000} \times 24 \times \text{Cost}$$

Ta = total amperage
V = voltage
24 = hours per day
Cost = cost per kilowatt hour
Cost/Day = total cost per day

On average, the cost per piece of drying equipment is \$1.00 to \$1.50 per day – and this rule of thumb can also be applied to answer inquiries. Ultimately, the building owner must contact the insurance company for direction when requesting reimbursement for electrical costs.

Managing Heat Produced by Equipment

All electrical devices produce heat as a by-product of using electricity.

Two Methods to Measure Heat

1) measuring dry bulb temperature (Fahrenheit or Celsius), which is similar to measuring relative humidity
2) measuring actual heat (British Thermal Units)

TIP

Drying equipment will operate more effectively if heat is properly controlled. Refrigerant dehumidifiers, especially, require a set temperature range, normally between 70°F and 90°F (21°C - 32°C)

When evaluating actual heat, the appropriate measurement is the **British Thermal Unit (BTU), or (calories or joules)**. A BTU is a specific unit of heat, much as grains are a specific measurement of water vapor.

For cooling, British Thermal Units are often expressed in increments of 12,000 BTUs per hour, or one "ton" of heating or air conditioning. This represents the amount of heat that must be added to melt a ton of ice at 32°F over 24 hours. The heat produced by drying equipment can be estimated.

Estimated BTUs

Amps x Volts x 3.4 = BTUs per Hour

Restorers can use the product of this forumala (BTUs) to size portable air conditioning units when they are necessary. It is important to note that most air conditioning units are rated in tons of air conditioning. Each ton represents 12,000 BTUs.

Most drying environments call for the use of mechanical dehumidification; therefore removing excess heat is often an issue. Desiccant dehumidifiers operate best in cool conditions, while refrigerants work best between 70° and 90°F (21°-32°C), so controlling heat is critical. Heat problems are most prevalent in warmer areas during summer, but temperature must always be monitored. Common ways to offset heat in structures include:

- Using the structure's air conditioning
- Installing portable air conditioning
- Using cool outdoor air (by setting an air mover on negative pressure to bring in cool, dry air)

When using outdoor air, restorers can keep the structure from getting too cool by running the air mover on a thermostat; this way, the unit turns off before the structure cools too much.

DEHUMIDIFICATION IN REVIEW:

Each dehumidification technology offers a unique set of performance and application advantages. Understanding the advantages of each technology is critical in deciding which dehumidifier will best manage the humidity in each restorative drying project.

> **Conventional Refrigerants and Low Grain Refrigerants (LGR):** Both perform best in humid environments where temperatures are above 70°F (21°C), as in situations that involve large amounts of wet porous materials (i.e., carpet, drywall). In all cases, LGR dehumidifiers will perform better than conventional refrigerants.
>
> **Desiccants:** Perform best in low humidity and when temperatures are cool, at or below 70°F (21°C). Water intrusions that will generate low humidity are those that involve primarily dense materials (i.e., hardwood, concrete).

When outdoor conditions are favorable, an open drying system can remove water more effectively than mechanical dehumidification. When an open system is possible, the restorer must consider outdoor humidity, temperature and building security.

For best results: The filtration system is a critical component of every mechanical dehumidifier. To maintain optimal performance, filters must be cleaned regularly and replaced as necessary. Restoration contractors need to read the equipment owner's manual or refer to other manufacturer's literature to understand how frequently filters should be cleaned or replaced.

If filter maintenance is not properly addressed, airflow through the dehumidifier will be restricted, compromising performance. Additionally, particulate that overloads the filter can lead to a buildup of dust, dirt and other foreign objects in the dehumidifier itself. This particulate will collect on surfaces in the machine and reduce the unit's performance dramatically.

DEHUMIDIFIER SAMPLE PROBLEMS

Dehumidifier Problem 1 – Imperial

Question: When using conventional refrigerant dehumidifiers in a Class 3 water loss containing 15,000 ft^3, what is the initial AHAM rated capacity required?

Solution:

- *Area:* 15,000 cu. ft.
- *Class:* 3
- *Factor:* 30 (conventional refrigerant and Class 3 – from the chart on page 100)
- *Math:* cu. ft. ÷ factor = AHAM capacity required
  15,000 ÷ 30 = 500 AHAM pints
- *Answer:* **500 AHAM pints**

Dehumidifier Problem 1 – Metric

Question: When using conventional refrigerant dehumidifiers in a Class 3 water loss containing 425 m^3, what is the initial AHAM rated capacity required?

Solution:

- *Area:* 425 cu. m.
- *Class:* 3
- *Factor:* 1.8 (conventional refrigerant and Class 3 – from the chart on page 100)
- *Math:* cu. m. ÷ factor = AHAM capacity required
  425 ÷ 1.8 = 236 AHAM liters
- *Answer:* **236 AHAM liters**

Dehumidifier Problem 2 - Imperial

Question: In a 80' x 40' structure with 8' ceilings, and using low grain refrigerant dehumidification, what is the AHAM capacity performance (in pints) required initially on an in-place restorative drying job with 12" of wicking up the walls?

Solution:

- *Area:* 25,600 cu. ft. (80 x 40 x 8)
- *Class:* 2 (in place and less than 24" of wicking)
- *Factor:* 50 (LGR and Class 2 – from the chart)
- *Math:* sq. ft. ÷ factor = AHAM capacity required
  25,600 ÷ 50 = 512 AHAM pints
- *Answer:* **512 AHAM pints are needed**

Dehumidifier Problem 2 - Metric

Question: In a 24 m x 12 m structure with 2.4 m ceilings, and using low grain refrigerant dehumidification, what is the AHAM capacity performance (in liters) required initially on an in-place restorative drying job with 30 cm of wicking up the walls?

Solution:

- *Area:* 691 cu. m. (24 x 12 x 2.4)
- *Class:* 2 (in place and less than 61 cm, of wicking)
- *Factor:* 3 (LGR and Class 2 – from the chart)
- *Math:* cu. m. ÷ factor = AHAM capacity required
  691 ÷ 3 = 230 AHAM liters
- *Answer:* **230 AHAM liters**

Dehumidifier Probelm 3 - Imperial

Question: When using desiccant dehumidification in a water loss with low permeance/porosity materials (plywood, concrete, no wet carpet/pad) in a 3,000 ft^2 restaurant with 10' ceiling height, what is the initial CFM required?

Solution:

- *Area:* 30,000 cu. ft. (3,000 x 10)
- *Class:* 1 ("low permeance/porosity...no wet carpet/pad")
- *Factor:* 1 (desiccant and Class 1 – from chart)
- *Math:* sq. ft. x factor ÷ 60 = CFM required
  30,000 x 1 ÷ 60 = 500 CFM
- *Answer:* **The initial equipment is 500 CFM.**

Dehumidifier Probelm 3 - Metric

Question: When using desiccant dehumidification in a water loss with low permeance/porosity materials (plywood, concrete, no wet carpet/pad) in a 279 m^2 restaurant with 3 m ceiling height, what is the initial CMH required?

Solution:

- *Area:* 837 sq. m. (279 x 3)
- *Class:* 1 ("low permeance/porosity...no wet carpet/underlay")
- *Factor:* 1 (desiccant and Class 1 – from chart)
- *Math:* cu. m. x factor = CMH required
  837 x 1 = 837 CMH
- *Answer:* **837 CMH**

Dehumidifier Problem 4 - Imperial

Question: When using 150 CFM desiccant dehumidifiers on a water loss where the water came from above and the job is being dried in place, how many dehumidifiers are needed for 16,500 cu. ft.?

Solution:

- *Area:* 16,500 cu. ft.
- *Class:* 3 (water came from above, in place)
- *Factor:* 3 (desiccant and Class 3 – from chart)
- *Math:* cu. ft. x factor ÷ 60 = CFM required
  16,500 x 3 ÷ 60 = 825 CFM
- *Answer:* 825 CFM are needed. Each desiccant dehumidifier will process 150 CFM, so **six desiccant dehumidifiers are needed on this job.**

Dehumidifier Problem 4 - Metric

Question: When using 255 CMH desiccant dehumidifiers on a water loss where the water came from above and the job is being dried in place, how many dehumidifiers are needed for 467 cu. m.?

Solution:

- *Area:* 467 cu. m.
- *Class:* 3 (water came from above, in place)
- *Factor:* 3 (desiccant and Class 3 – from chart)
- *Math:* cu. m. x factor = CMH required
  467 x 3 = 1401 CMH
- *Answer:* 1401 CMH are needed. Each desiccant will process 255 CMH, so **six desiccant dehumidifiers are needed on this job.**

Case Study

A structure is 2,000 sq. ft. (186 sq. m.) with an average 9' (2.7 m) ceiling. Water is wicked 8" to 14" (20-36 cm) on many walls in the affected area, and 1,400 sq. ft. (130 m^2) is affected. The source of the water was clean and the job will be dried in place. Rooms affected include:

- Bedroom, den and hallway with berber carpet
- Bathroom with vinyl flooring that will be removed
- Foyer, dining room and kitchen with hardwood floors

The unaffected area can be separated from the affected area by closing doors. And, ordinary contents are present in the structure.

Question 1: (Imperial)

How many 100 pint LGR dehumidifiers should be used on this loss?

Solution 1:

- *Area:* 1,400 x 9 = 12,600 cu. ft. or
- *Class:* 2 (wet carpet with wicking less than 24")
- *Factor:* 50
- *Math:* 12,600 ÷ 50 = 252 AHAM pints needed
- *Answer:* **Three 100-pint LGRs are needed.**

Question 2: (Imperial)

How many air movers should be brought to this job initially, to ensure airflow across all wet surfaces?

Solution 2:

- *Area:* 1,400 sq. ft.
- *Math:* 1,400 ÷ 60 = 23
  and 1,400 ÷ 50 = 28
- *Answer:* **23-28 air movers are needed.**

Question 3: (Imperial)

How many 500 CFM AFDs are needed on this job?

Solution 3:

- *Area:* 12,600 cu. ft.
- *Factors:* 4-6 air exchanges per hour
- *Math:* 12,600 x 4 ÷ 60 = 840 CFM
  and
  12,600 x 6 ÷ 60 = 1,260 CFM
- *Answer:* Between 840 CFM and 1,260 CFM are needed. Therefore, **2-3 AFDs are needed.**

Question 1: (Metric)

How many 47 liter LGR dehumidifiers should be used on this loss?

Solution 1:

- *Area:* 186 x 2.7 = 502 cu. m.
- *Class:* 2 *(wet carpet with wicking less than 61 cm)*
- *Factor:* 3
- *Math:* 502 ÷ 3 = 167 AHAM liters needed
- *Answer:* **Four 47-liter LGRs are needed**

Question 2: (Metric)

How many air movers should be brought to this job initially, to ensure airflow across all wet surfaces?

Solution 2:

- *Area:* 130 sq. m.
- *Math:* 130 ÷ 5.6 = 23
  and 130 ÷ 4.6 = 28
- *Answer:* **23-28 air movers are needed**

Question 3: (Metric)

How many 850 CMH AFDs are needed on this job?

Solution 3:

- *Area:* 502 cu. m.
- *Factors:* 4-6 air exchanges per hour
- *Math:* 502 x 4 = 2008 CMH
  and 502 x 6 = 3012 CMH
- *Answer:* Between 2008 and 3012 CMH are needed. Therefore, **2-4 AFDs are needed.**

Chapter 9

Drying Systems

IN THIS SECTION

INTRODUCTION

Restorative drying contractors who use the most efficient drying systems will succeed because they save money for their customers, make money from their efforts, and reduce liability on every drying job. The best professionals manipulate and control as many factors as possible in their drying system to finish each job as soon as practical, including:

- Extracting as much water as possible
- Lowering humidity in the affected air
- Increasing airflow across wet surfaces
- Controlling temperature of air and surfaces
- Using outside air when possible
- Creating a drying chamber
- Adjusting or removing equipment when necessary
- Removing items that should not be dried
- Saving items that can be dried in place
- Using specialized tools to dry difficult areas

Additionally, knowledgeable restoration contractors recognize that each loss site is different, and they adjust methods based on each situation. They recognize that what was done on the last job may not be the best method for the current job, or the next one.

As the industry changes, these principles will remain much the same. Each new technology offers another way in which to perform these tasks more efficiently, more effectively and with a higher rate of success.

TIP

Several systems can be used to control humidity and temperature in a drying environment. Choosing the appropriate system is a critical component to successful structure drying.

CLASSES OF WATER: RATE OF EVAPORATION

Each water intrusion involves different materials, a different source of water, and a different quantity of moisture – and each situation requires a unique and specific equipment set-up. The key when determining how much equipment to place initially is in knowing the potential evaporation rate. If evaporation rates will be slow, less equipment will be needed to remove the boundary layers and moisture from wet surfaces. On the other hand, if evaporation is rapid, more dehumidification and air movement will be needed to deal with the moisture being removed from the wet materials.

A wet structure will often contain areas of both rapid and slow evaporation, presenting a variety of drying challenges. When this is the case, a variety of systems are employed, each tailored to a specific challenge.

A tool used to determine equipment needs is water classification. Classes of water describe the potential rate of evaporation based upon the amount of water present and the type of material affected. Equipment types and quantities are then determined based upon the potential rate of evaporation defined by each class.

Class 1

With a **Class 1** water intrusion, evaporation will progress at a slow rate because there is minimal water present in the structure. Class 1 intrusions involve only a part of an area, or involve a larger area that has only been lightly affected. In a Class 1 loss, there is little or no wet carpet or cushion; typically, only low porosity materials (such as plywood, concrete and structural lumber) are affected. Some examples of common Class 1 water intrusions are:

- A concrete basement floor that only absorbed a small amount of water.
- A structure that was recently affected by sewage and all wet porous materials were removed.
- Losses where carpet and pad have been removed and there is no wet gypsum.

In a Class 1 intrusion, the moisture laden boundary layer will be minimal because there is a slow rate of evaporation. Therefore, Class 1 intrusions require the least amount of equipment. Since evaporation is slow, less dehumidification is needed to keep a balanced drying system.

Class 2

With a **Class 2** water intrusion, evaporation will progress at a faster rate because there is a large amount of water present in the structure. Class 2 water intrusions involve damage to entire areas of carpet and pad. Water may have wicked up the walls but is less than 24" high, and structural materials are wet. Porous and low porosity materials are affected. Some examples of common Class 2 intrusions are:

- A loss that includes wet carpet, underlay (pad, cushion) and gypsum wicked less than 24" (61 cm).
- A structure that was affected by Category 2 water (gray) and where the underlay (pad, cushion) was removed, but wet carpet, drywall and structural materials remain.
- Structures with no wicking on walls but where the carpet and underlay (pad, cushion) is being dried in-place.

In a Class 2 intrusion, moisture laden boundary layers will be present around all wet porous materials because of the fast rate of evaporation. Therefore, Class 2 losses require more equipment than Class 1. Since evaporation is fast, more dehumidification is needed to keep a balanced drying system.

Class 3

With a **Class 3** water intrusion, more water is present than with any other class; therefore, evaporation progresses the fastest. Class 3 water intrusions involve damages to entire areas of carpet and underlay (pad, cushion). Moisture has wicked up the walls more than 24" (61 cm), water may have come from above, and/or wet insulation may be present. There are wet structural materials present, along with great amounts of wet porous items. Some examples of class 3 losses are:

- Strong winds damage the roof and rain enters from above, saturating ceilings, walls, flooring materials and structural items.
- Second floor water supply failure saturates

entire areas below with large amounts of water (second floor could be Class 1 while the main floor would be Class 3).

- Overhead water supply pipes malfunction, saturating entire areas with large amounts of water.

In a Class 3 intrusion, moisture laden boundary layers are present around all wet porous materials because of the fast evaporation and amount of water present. Class 3 water intrusions require the most drying equipment of all; more dehumidification is needed to ensure a balanced drying system.

Class 4

A **Class 4** water intrusion results in a specialty drying situation. Typically, wet materials present in a Class 4 situation have very low porosity. Water has saturated deeply into the materials and will require very dry air to return to pre-loss condition.

Materials that are common to Class 4 intrusions include hardwood, plaster, brick, deeply saturated concrete, deeply saturated ground soil and stone.

EXAMPLES OF CLASS 4 INTRUSIONS

- Plaster and lathe walls that are deeply saturated
- Gymnasium floors
- Hardwood floors in residential construction
- Very old construction with multiple layers of building materials
- Concrete
- Dirt floors

Use of desiccant and/or LGR dehumidification is necessary in Class 4 situations. Lower GPP, and therefore lower vapor pressure, is needed to pull moisture out of the wet materials. Because of the slow rate of evaporation in these losses, there will be little moisture present in the boundary layer. Less air movement is needed than in other classes of intrusion. It is used only to facilitate circulation of dry air, and not to sweep away saturated air (boundary layers) at the surface of materials.

OPEN, COMBINATION, AND CLOSED DRYING

Ultimately, the goal in restorative drying is to facilitate a rapid rate of evaporation while preventing the spread of moisture to unaffected materials. In order to accomplish this, dehumidification must occur at a rate equal to or greater than the rate of evaporation. Dehumidification can take multiple forms: either *mechanical dehumidification* or the *exchange of drier air from another location to replace humid air in the affected area.* This presents three options for an overall dehumidification system:

- **Opcn** – using outdoor air to lower indoor humidity
- **Closed** – using mechanical dehumidification only
- **Combination** – using some outdoor air combined with mechanical dehumidification

OPEN, CLOSED OR COMBINATION?

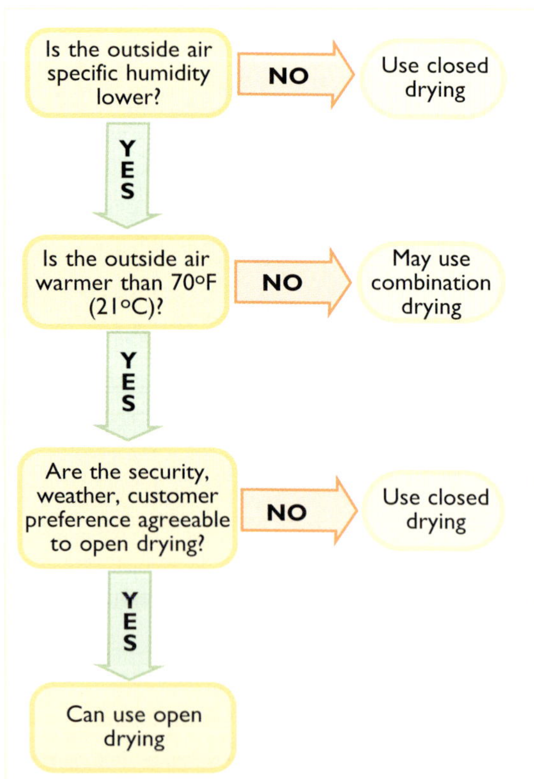

Choosing the appropriate drying system.

Deciding which dehumidification system to use is a simple process.

- First, restorers compare the indoor air and outside air. If the air outside is wetter (higher GPP), a closed drying system is used.
- Second, restorers consider air temperature. When outside air is below 70°F (21°C), a closed system is used, or additional heat is used in conjunction with an open or combination drying system.
- Third, restorers consider additional factors, such as building security, changing weather, customer preference or other factors that will not allow open drying systems to be used.

Regardless of the system used, the environment is monitored frequently to ensure that the indoor humidity and temperature are conducive to evaporation. As the environment dries and as conditions change, the dehumidification system is modified as needed. Additional information on how to evaluate changing conditions can be found in *Chapter 4 Inspection, Monitoring and Documentation* and *Chapter 7 Drying Science.*

Using Open Drying Systems

When outdoor conditions are favorable for open drying systems, several options are available. Although, outdoor air may be beneficial because of its lower humidity (GPP), outdoor air is less favorable if it is too cold to support evaporation. Therefore, additional equipment may be necessary to control indoor temperatures while using outdoor air.

If open drying is determined appropriate, the following options (while keeping building security in mind) are available to the restorer:

- Using ventilation fans installed in the structure
- Opening windows slightly
- Using fans to create pressure differentials and air exchange
- Opening fireplace flues *(Note: When opening fireplace flues, it is important to ensure that negative pressure is not pulling air the wrong direction through the flue.)*

A more aggressive open air drying system is the air exchange heat drying system. Air exchange heat drying provides a way to use dry outdoor air while keeping the temperature inside the structure at a level conducive to rapid evaporation. This system increases the temperature of the outside air, reducing relative humidity and increasing thirst. (It is important to note that the humidity ratio, or GPP, of the outside air remains the same when heat is added.)

In an air exchange heat drying system, the heated outdoor air is circulated throughout the structure and across wet surfaces, promoting evaporation. The air becomes wetter as it takes on evaporating water, and is eventually vented out of the structure. A continuous re-supply of outdoor air is heated to replace the exhausted, moisture-laden air.

When this type of system is employed, ventilating in a manner which supports a strong negative pressure is important to prevent damage. If a positive pressure is allowed to develop, the warm air and evaporated water may travel into unaffected areas of the structure and into interstitial cavities (e.g., wall cavities, beneath and behind cabinets). If warm, relatively dry air is allowed to collect near cool, unaffected surfaces, the air's temperature will decrease. This will result in an elevated relative humidity capable of causing secondary damage.

Many methods of monitoring are especially important when using air exchange heat drying systems. As water is evaporated through the application of heat, it will leave building materials in virtually every vapor direction. This will result in evaporation towards the hot, dry air – but will also yield evaporation in the opposite direction.

For example, if heat were applied to a wet plywood subfloor by blowing hot air over the top surface, water vapor would evaporate directly into the hot, relatively dry air stream. This water vapor would then be carried to an exhaust and evacuated from the structure.

Drying Systems

Water vapor would *also* evaporate and exit the bottom of the subfloor. Water vapor leaving this opposite side of the subfloor can cause secondary damage if not addressed. If it evaporates into a cool, unventilated crawlspace, it will cause elevated humidity and potential condensation. If it evaporates into a ceiling cavity (when drying a second floor), humidity will build up, causing the same end result.

This potential problem calls for the extensive use of thermo-hygrometers, deep probe moisture meters and a thorough understanding of the building's design. If these areas are not thoroughly monitored, hidden moisture will be left unaddressed.

For these reasons, air exchange heat drying systems are only used when the resultant humidity can be controlled and regulated to prevent high dew points, and when the restorer has complete access to all surfaces of affected materials. Open drying and heat drying effectiveness is monitored by using a system similar to the grain depression method used when monitoring mechanical dehumidifiers.

Monitoring Open Drying Systems

- Monitor the outdoor GPP and compare it to the GPP of the air being removed from the structure. If the system is generating evaporation, the air exiting the structure should be wetter than the outdoor air entering.
- Use the difference in GPP as the grain depression or water removal in grains per pound of dry air.
- Finally, use the mathematical formula discussed in *Chapter 7 Drying Science* to convert the grain depression to pints of water removal in a 24-hour period.

Using Closed Drying Systems

A closed drying system is the traditional method of drying when outdoor conditions are wetter than indoor conditions, or when conditions are generally not favorable for open or combination drying. Closed drying involves the use of mechanical dehumidification inside – while keeping wet air outside.

There are several reasons to choose a closed drying system. When mechanical dehumidification is used in a closed system:

Benefits of a Closed Drying System

- The restorer has a high level of control.
- Outdoor weather changes will have minimal effect on indoor conditions.
- Indoor humidity is regulated to prevent high indoor dew points.

Additionally, water vapor is encouraged – through the use of low vapor pressure – to move predominantly toward the indoor air mass (as opposed to sub-surfaces and exterior cavities). This is especially true when outdoor humidity is high. Using an open system, the restorer is *not* capable of achieving a vapor pressure any lower than that of the outdoor air. Any evaporation will result in vapor pressure higher than the outdoor air.

Effectiveness of closed drying is improved when restorers use the following principles:

- Ensure that there are no pressure differentials between the outside and inside air. Both negative and positive pressures will result in an exchange with humid air from outside.
- Encourage customers to keep all windows and doors closed at all times.
- Turn off all bath vents (extraction fans), as bath vents create a negative pressure and will cause infiltration of wet outside air.

Effectiveness of a closed drying system is monitored by evaluating dehumidifier grain depression, and by tracking the change in indoor humidity (GPP). Humidity will noticeably decrease over time, indicating that the dehumidification is removing water vapor at a greater rate than evaporation is adding water vapor to the air.

During the initial stages of a closed system, it is especially important to monitor indoor humidity. Typically, the indoor humidity will be exceptionally high as evaporation rates will be very rapid. There may be a need, even in warm and humid geographical regions, for a temporary use of outdoor air. Temporary use of outdoor air is considered whenever indoor humidity (GPP) is higher than outdoor humidity.

Using Combination Drying Systems

Combination drying systems merge the benefits of both open and closed drying: the use of available, dry air from outside *and* the control offered by a closed system and mechanical dehumidification. There are three ways to set up a combination drying system: temporary flushing, slow and continuous ventilation, and strong and continuous ventilation.

Temporary Flushing. If outside air is significantly drier than indoor air, restorers can open windows and doors for a short time. This can effectively reduce indoor GPP. Indoor conditions are monitored with a thermo-hygrometer to see when humidity levels approach equilibrium with outdoor levels. Temperature loss is also closely monitored to prevent cooling of the indoor environment – which will inhibit evaporation.

The temporary use of outdoor air is halted when indoor and outdoor humidity (GPP) approach equilibrium, or when indoor temperatures fall below 70°F, whichever occurs first. If the temperature falls below 70°F, yet indoor humidity (GPP) remains higher than outdoor humidity, the exchange of air can be slowed to allow for continued flushing while preserving indoor air temperatures. If needed, the process can also be accelerated by temporarily setting up negative pressure with an air mover.

Slow, Continuous Ventilation. Most structures have small ventilation fans (e.g., bath fans or range vents) that can be used to set up a slight negative pressure inside a structure. Small, household ventilation fans move 100-200 CFM (170-340 CMH), and this air exchange can have a significant effect on the humidity level in the structure without a negative impact on indoor temperature. However, whole house ventilation fans (1,000-2,000 CFM or 1,700-3,400 CMH) are not appropriate for this use because no temperature control is established and high levels of ventilation can dramatically drop indoor air temperatures, inhibiting evaporation.

Ventilation fans used must ventilate to the outside of the structure, not into an attic or crawlspace. Restorers can prove the effectiveness of this system by comparing the vented air (same GPP as affected area) to the outside air GPP. They can then factor this "grain depression" with the CFM of the vent fans to calculate water removal, (see *Chapter 7 Drying Science).*

Strong, Continuous Ventilation. Restorers can generate a strong negative pressure using a carpet dryer, high pressure axial fan, or similar air mover. The device must be thermostatically controlled to ensure that air exchange with the drier outside air only occurs when conditions are appropriate.

This type of system is set up by ducting an air mover to blow out a secured window. The air mover is plugged into a thermostat-controlled power source, so that the air mover is only on when the temperature inside the structure is above a certain point, usually 70-75°F (21-24°C). The result is a repeating cycle of an open (flushing) and closed system.

When conditions are appropriate, a combination system is often the most cost effective for the customer. Monitoring a combination system requires tracking grain depression on both the mechanical dehumidification systems and the open air exchange. Each is evaluated using the formula in *Chapter 7 Drying Science*, which converts grain depression to water removal in pints per 24 hours.

Drying Systems

When flushing is temporary, the calculation must be adjusted. For example, if a strong, continuous ventilation system is used on a thermostat, the amount of time the exhaust is active must be monitored. If it is active for 20 minutes per hour, the total pints removed would be 33%, or one third, of the total pints calculated using the grain depression formula in *Chapter 7 Drying Science.*

DRYING CHAMBERS

In most restorative drying projects, the entire structure is not affected – and unaffected areas may need to be protected from variable humidity. Also, the effectiveness of mechanical dehumidification may need to be concentrated or focused.

Before drying is complete, the affected area will often contain drier air than unaffected areas. Thus, higher humidity may need to be prevented from mixing with this dry air in the affected area.

These goals can be accomplished by creating a drying chamber. A drying chamber is defined as *the area in which the restorer will manipulate and control temperature, humidity and airflow to promote evaporation.* Drying chambers can be set up in a number of ways, including:

- Closing doors
- Taping up plastic sheeting
- Erecting semi-permanent barriers with plastic sheeting and expanding poles, shower rods, etc.

If barriers are set up between the affected and unaffected areas, restorers must be sure to keep practical factors in mind – such as *customer and employee access* and *pressure differentials.* **Pressure differentials** will occur whenever air is added to or removed from the affected area – for example, when the structure's HVAC system is running. This is particularly a concern if the system contains only one HVAC return (the intake for the system, drawing air from the structure and returning it to the HVAC mechanical system).

In the area where the HVAC return is, there will be a negative pressure. Restorers must consider the influence of negative pressures whenever attempting to establish a drying chamber, and evaluate whether or not it is a concern. There is also a possibility of spreading contaminants, such as humidity, odors, and particulates whenever pressure differentials exist.

Although it could be beneficial to close doors that separate the affected and unaffected areas, there are also cases where it is necessary to remove doors in the affected area from their hinges. This practice prevents the restriction of airflow along affected walls and allows for good air circulation throughout the affected area. It is especially important to remove doors if walls immediately behind them are wet. If doors close off a portion of the affected area, either leave them open or remove them; a closed door will cause an unbalanced drying system and potential secondary damage (from lack of dehumidification). When doors are removed, they are labeled for easy resetting at the end of the job.

CHOOSING A DRYING METHOD

Drying science has advanced to the point where nearly any material item in a structure can be dried. There are many tools and techniques to achieve drying goals, and a true professional will have all of these tools at their disposal. Since everything can be dried, the critical question in restorative drying is: *"Should it be dried?"*

Factors that influence drying method

When determining if a wet material or item should be dried, restorers consider three factors:

- Contamination
- Damage
- Cost

CONTAMINATION
DAMAGE
COST

Contamination must always be the first consideration when determining if a material should be dried. Porous items that are affected by Category 3 water, for example, must be removed. Only high value porous contents can be saved after a Category 3 water intrusion. Even in a structure contaminated by Category 2 water, highly porous items, especially underlay (pad, cushion) beneath carpet, must be removed.

Good professional judgment is important when evaluating contaminated environments. The health and safety of the customer and employee are the primary concern. If an item cannot be economically returned to pre-loss cleanliness or better, it is removed and disposed of. (More information on contamination and categories of water can be found in *Chapter 10 Microbiology.*)

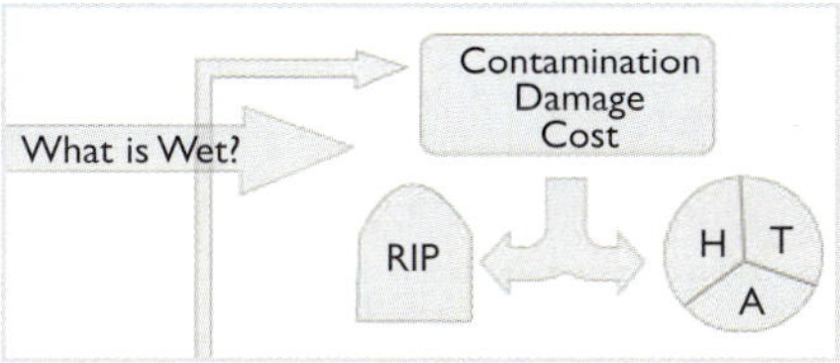

Damage is also an important consideration when choosing a drying method. If a material or item is damaged beyond economical repair, it is removed from the structure. For example, fallen ceilings, insulation that is packed down, and particle board furniture that is beyond repair are all discarded. Drying will not return these permanently damaged materials to a pre-loss condition, and additional repairs would be necessary after drying is complete. Restorers evaluate the cost of repair and the cost of drying, compared to the cost of simply replacing materials.

Disposing of damaged materials early in the restoration process prevents incurring costs from both drying and replacing the material. When damaged materials are removed, water absorbed by the material is also removed. This speeds the drying rate for the rest of the affected materials. Leaving wet, permanently damaged material in the environment will increase the need for dehumidification and airflow.

Cost is the final consideration in choosing a drying method. It must cost less to dry an item than to replace it. For example, a base cabinet installation that is affected by water may cost $1,500 to replace. Therefore, drying the cabinet must cost less than $1,500 in order to be practical. Costs evaluated must include replacement *and* historical value, additional wiring and business interruption.

Many restorers aim for a drying cost of at least 50% less than replacement. When estimated costs for drying break the 50% threshold, the restorer communicates openly with materially interested parties to ensure that the customer is involved in the decision-making process. Whenever estimated drying costs exceed replacement costs, restorers must thoroughly document the outcome of this communication and the final decision.

"Should it be dried?" is the primary question. If restoration is not supported after evaluating the level of contamination, damage and costs, then the items are removed. On the other hand, if restoration is supported, the item is dried.

Two Drying Methods

After making the decision of whether or not to dry a material, restorers begin evaluating how to begin the restoration work. There are many methods available – each applying a different combination of humidity control, temperature, airflow and physical manipulation of the material (e.g., injection of airflow, perforation, removing finish materials).

Restorers use the information obtained during evaluation of materials to help select the best drying method for the job. Generally, there are two primary methods to promote drying of affected structures: 1) *disruptive* methods and 2) *aggressive* methods.

Disruptive Drying Methods

Disruptive drying methods involve removing wet items, injecting air to speed drying, or perforating surfaces to allow water to evaporate. The term *dis-*

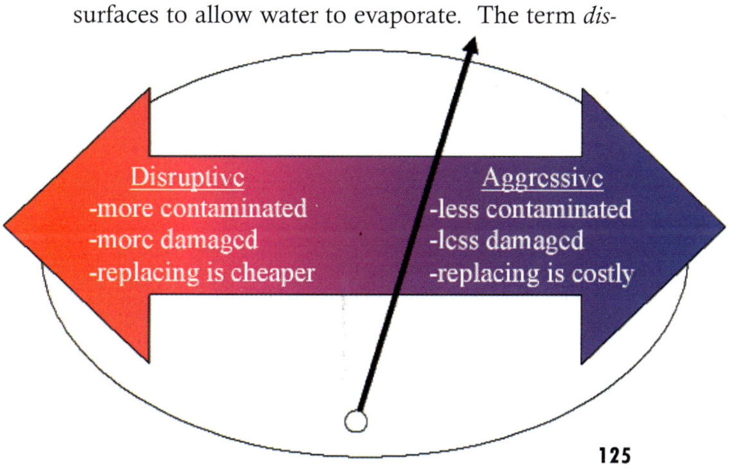

ruptive is used because repairs will have to be done after the structure has been dried. Use disruptive methods when contamination, damage, cost or customer concerns require removal or manipulation of the affected material.

Disruptive Methods

- ☑ Removing underlay (pad, cushion) after Category 2 water intrusion
- ☑ Removing damaged baseboard
- ☑ Removing vinyl flooring to promote drying of underlayment
- ☑ Perforating vinyl wallpaper
- ☑ Injecting air into wall cavities with an inter-air drying system
- ☑ Cutting out wallboard to remove wet blown-in insulation
- ☑ Disposing of damaged particle board furniture

Disruptive methods are required in many situations where contamination or damage is present, or when it is simply not cost-effective to dry an item. As mentioned earlier, whenever these methods are employed it is important to evaluate the cost of replacement and repair. Materials of high value are generally dried, not replaced, even if damage is present.

Aggressive Drying Methods

Aggressive or "in place" drying methods involve leaving wet items in the structure and drying them in place using warm, dry direct airflow. Aggressive methods are used when contamination and damage are not concerns, and when it is cost effective to dry an item instead of replacing it.

Restorers use aggressive drying methods when all the following are true:

- The water intrusion came from a sanitary source (Category 1).
- The structure has been wet for less than 72 hours.
- Drying carpet and underlay (pad, cushion) in place will not cause damage to subfloor (especially hardwood).
- Adequate dehumidification is available and usable on site.
- Deep extraction tools are available.

When aggressive drying is used, the water intrusion will be *classified* as a Class 2 or Class 3 because of the additional moisture being left in the environment.

Which Method Is "Correct"?

A great amount of restorers' energy has been spent debating which method of drying – *disruptive* vs. *in place (aggressive)* – is the correct way to dry a structure. In fact, both methods are viable ways to reach the desired goal: *returning a structure to pre-loss condition*. Further, in nearly every structural drying project, *both* methods are used. The following are common ways in which both methods are typically employed:

- Wet underlay (pad, cushion) is removed, but carpet is dried.
- Baseboard is removed, but wallboard is dried.
- Vinyl flooring is removed, but wet underlayment is dried.

Even in a disruptive drying scenario, some items are dried in place. This concept can also be illustrated by considering an aggressive drying scenario. Even when the majority of materials are dried in place, several materials may be physically altered or removed. Some examples may include:

- Vinyl flooring removed to expose wet underlay
- Baseboard removed to access wall cavity, drying wall assembly in place
- Kick plates removed to blow dry air beneath a cabinet, which is left in place

In other words, even within an in-place drying scenario, some items may be removed. Each restorative drying job requires an appropriate balance of disruptive and aggressive methods. Where each job ends up on the disruptive or aggressive scale depends upon the amount of contamination and damage present, and the cost effectiveness of drying wet items. The ultimate goal is to prescribe the most cost-effective combination of each method, uniquely balanced on a case by case basis.

A good general practice during the initial inspection process is to assume an aggressive structural drying approach. Restorers diligently search for any obstacles to this aggressive approach, and modify the aggressive process accordingly. Wherever obstacles are absent, the aggressive approach remains the method of choice.

After a thorough initial inspection, restorers install equipment according to the formulas and techniques discussed in *Chapter 8 Drying Equipment*. They then inspect every affected material and document the amount of moisture in accordance with the procedures outlined in *Chapter 4 Inspection, Monitoring, and Documentation.*

After 24 hours (or less) of drying, materials are re-inspected (monitored) to determine if progress was made. As each material's progress is documented, equipment and processes are adjusted as follows:

- **Material Is Drying** – If the material is drying, continue with the current drying approach. It is working properly and progress is being made.
- **Material Has Dried** – If the material has achieved its drying goal, reposition equipment to focus on other areas, or remove it from the structure.
- **Material Is Not Drying** – If the material is not making progress, modify the approach. It is not working. Modification may be made to the equipment or the material, depending upon which is most economical.

The absolute constant in a wet structure is *change.* Change, when controlled closely, is a positive change: a change toward pre-loss condition. If the desired change is not evident (e.g., drying progress), it is necessary to control the change by other means. At this point, it is critical to re-evaluate the material:

- **Contamination** – Now that it has been wet for another 24 hours, is there a concern for contamination? Can this contamination, if present, be treated properly? Should the material be removed to ensure complete removal of the contaminant, if present?
- **Damage** – Now that the material has been wet for another 24 hours, has permanent secondary damage become evident, such as warping, swelling, splitting or staining? Can the damage be restored, cost effectively?
- **Cost** – The current system is not working and must be changed. What will result in the lowest cost – removal and replacement or more aggressive drying (e.g., more equipment)?

Using this approach of assessment and ongoing re-evaluation ensures that the structure is being thoroughly and properly dried for the lowest possible cost. Restoration will be maximized, reducing the amount of time necessary for repairs and reconstruction. Customers will regain access and normal occupancy of their structure as soon as possible. Insurers will close their claim files quickly. The documentation that results from the thorough, daily monitoring will also ensure that the file remains closed, and that "come back" claims will not haunt the restorer, the customer or the insurer.

Another benefit to this approach is the minimal labor necessary during the emergency response. Labor rates are typically much higher during the initial response, which is often performed during non-business hours. When this approach is applied, materials are only removed during the initial response if they show obvious signs of being difficult to restore.

Additionally, when materials are removed, the reason for removal is very apparent and can be thoroughly documented. Materials are removed only when damage or contamination dictates, when cost limitations are obvious, when customers mandate, or when drying progress is not evident in the first 24 hours.

DRYING PROCEDURES

With the variety of installations in a structure, it can be confusing to consider what will be a good approach to drying a material or surface. In these cases, refer to *Chapter 7 Drying Science* for guidance.

The basic rules of psychrometry always apply: *wet items will dry faster if drier air is supplied, with adequate airflow, at a good drying temperature.* Restorers can rely on this principle when drying challenges are present.

Drying Systems

Drying Walls

With a variety of surfaces, wet walls present a significant challenge to restorers. For basic walls with a single layer of wallboard and flat paint, the only drying tool needed is an air mover blowing warm, dry air along the wall. In some structures, the surface of the wall is covered with a non-permeable surface or there are multiple layers of wallboard. In these cases, it may be necessary to remove base molding, drill holes and introduce air movement. Airflow in the wall cavity increases the rate of evaporation and allows the wall to dry from both inside and out.

When ceilings are affected, drying procedures are similar to wall drying. If the ceiling is sagging or otherwise damaged, wet wallboard is removed and discarded. This not only removes water from the structure and speeds drying, but it also eliminates a potential safety hazard: wet, sagging ceilings can easily fall and cause injury.

When drying a wall cavity, restorers can follow some of these steps to ensure success:

- Determine the type of wall. This can be done by opening receptacle and switch covers, making inspection holes or by other intrusive means.
- Check for wicking with a non-invasive meter.
- Check insulation with an invasive, deep wall probe if the water came from above or standing water was above the bottom plate.
- Use the "dot system" (described in *Chapter 4 Inspection, Monitoring and Documentation*), an effective way to evaluate wall drying progress.

TIP

- Avoid damage to wall surfaces. Painting and repairing walls is expensive and time consuming.
- Generally, use more aggressive methods first. Employ disruptive drying only if no progress is being made in the first 24 hours.

Common Wall Assemblies

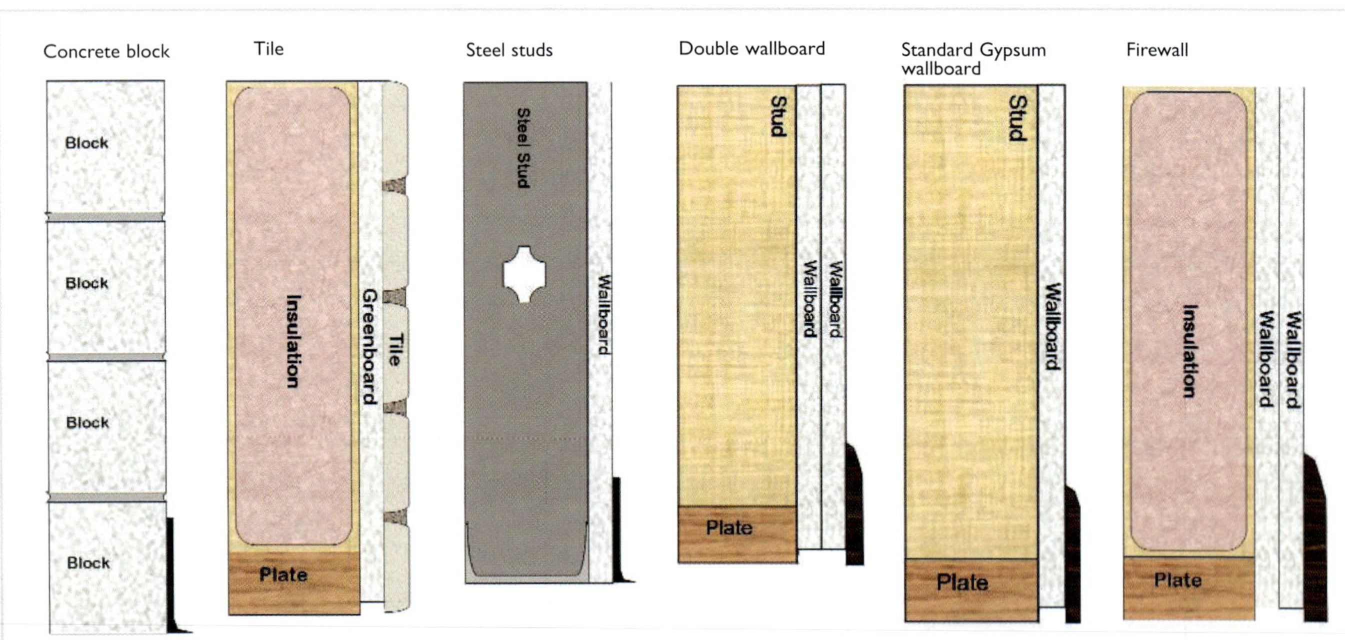

If a wet wall is not making progress after 24 hours of drying, restorers evaluate their options for modifying the drying method by considering the basic rule in regard to evaporation: *evaporation will increase as materials are made warmer, as air is made drier and as air is moved more rapidly across the wet surface.* Under this principle, there are several options for action, beginning with the least disruptive:

- Add more direct airflow on the outside of the wall.
- Move the dehumidifier output so that it is closer to the wet surface.
- Carefully remove baseboard and continue airflow along outside of wall.
- Drill holes below baseboard level to allow wall to breathe and continue airflow along outside of wall.
- Blow direct airflow into the wall using air movers or inter-air dryers.
- Feed air movers that are blowing into wall with a dehumidifier.
- Warm wall surfaces with heat lamps.
- Completely remove all wet wallboard and all wet insulation.

Clearly, each drying challenge is different. The most successful restorers will be ready to employ any and all of these options to reach drying goals efficiently.

Drying Floors

Drying technology can answer the challenge of nearly every flooring system. As with other structural and contents items, the question is not whether a floor can be dried, but: "*Should it be dried?*" Considering contamination, damage and cost-effectiveness is vital when drying floor systems.

Contamination in flooring must be viewed not only from the surface material, but also from sub-surfaces. For example, a hard floor surface can be cleaned after a Category 3 loss, but what if it is installed over a sleeper system? Is there a way to remove the contamination from beneath the floor and between the sleepers? If not, removal of the floor will be necessary.

Damage in flooring is seen most often in engineered wood ("was wood") or in flooring that has been exposed to water for long periods of time. The swelling of the material will cause damage to the material itself (especially in engineered lumber) or to the surrounding structure (as in hardwood pushing a wall off its base). Damage is common in the following flooring materials:

- Medium Density Fiberboard (MDF)
- Particle board
- Laminate flooring
- Parquet floor

Hardwood on plywood subfloor

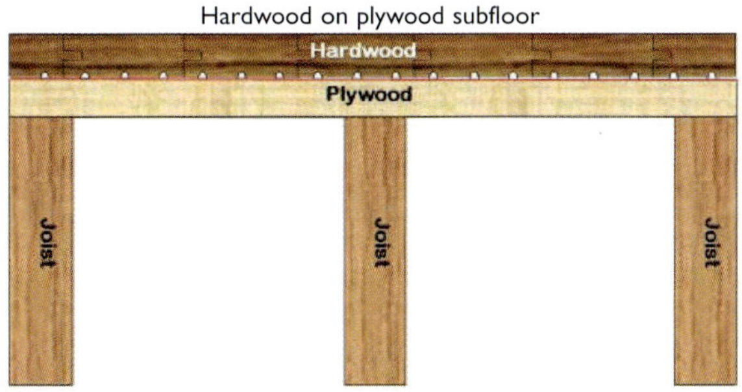

Tile on subfloor

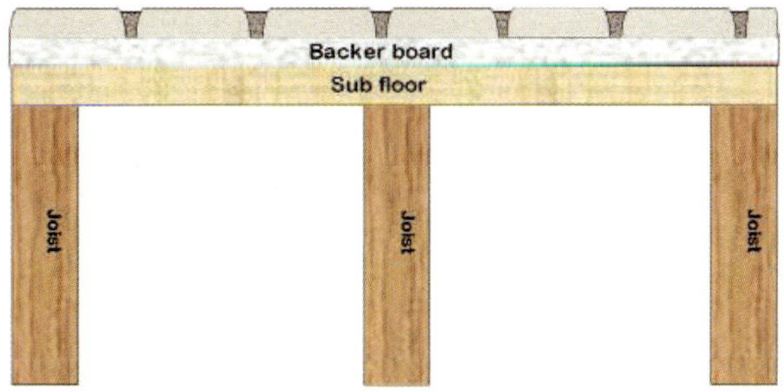

Standard carpet installation over subfloor

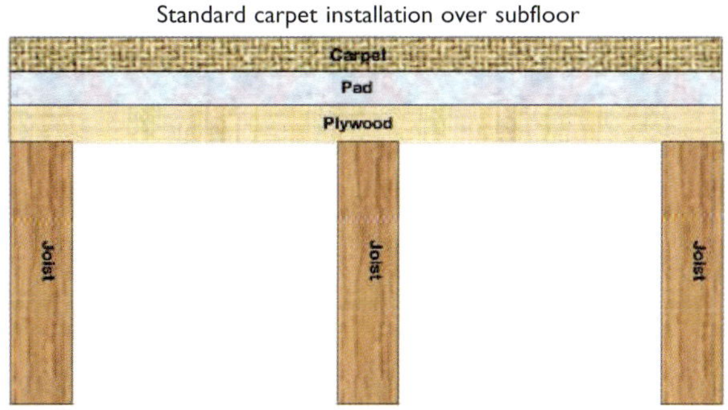

Standard carpet installation over concrete

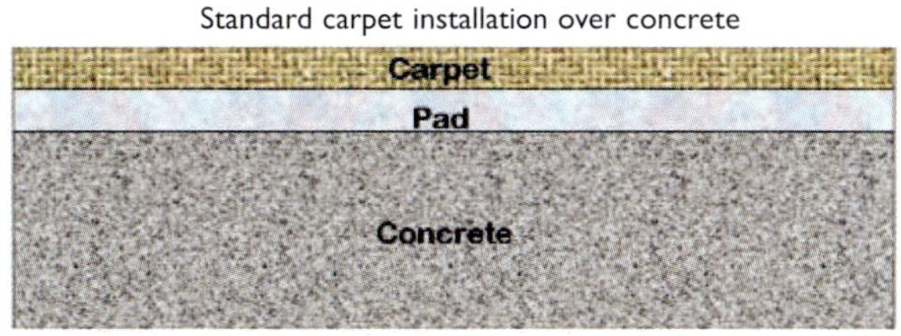

Additionally, *any* hardwood material that absorbs moisture will swell. The effect of each plank of a hardwood floor achieving its potential dimensional change, gaining 1/8" (3 mm) of width, is quite awesome. Buckled, heaving floors and walls pushed out of the way – physically moving the wall plate – are not uncommon in these situations. Permanently damaged flooring is removed promptly. More information on how hardwood flooring and other engineered materials respond to moisture is available in *Chapter 6 Materials.*

Cost of drying vs. cost of replacing is the final consideration in floor drying. Multiple layers of flooring may be present – and drying each layer in place may dramatically increase cost *and* time. It is important to closely evaluate this impact to ensure that restoration is not more expensive and time consuming than replacement. Some examples of floor drying challenges and the generally accepted resolutions are:

- **Wet carpet and underlay (pad, cushion) installed over hardwood flooring.** Discard underlay and remove carpet from the structure for drying off-site or for replacement after drying is complete.

- **Wet underlayment under vinyl flooring.** Remove the vinyl and discard it if underlayment is saturated and/or damaged; vinyl is a vapor barrier and time required to dry underlayment will be dramatically reduced.

- **Tile flooring over saturated underlayment, over vapor barrier, over plywood subfloor.** In most cases, remove the tile (because of the multiple layers). The only exception: irreplaceable or very expensive tile.

- **Multiple Finishing Materials, or "archeological digs": carpet and underlay (pad, cushion), over old carpet and underlay, over vinyl, over hardwood, over particle board, over shiplap...etc.** In most cases, remove all covering materials to expose the original subfloor, saving only the structural components.

The easiest way to determine what type of flooring is present is to remove HVAC floor registers, if present. This allows easy viewing of each layer of flooring. If no floor registers are present, materials are inspected by disengaging carpeting using a knee kicker, then using a carpet awl to lift carpet. Where carpet and floor registers are not available, other means will be necessary. On hardwood flooring, restorers can remove quarter-round molding or other trim and inspect from the side. If no access can be found to determine what floor system is being dealt with, it may be necessary to cut into materials in an inconspicuous area.

When drying a floor system, restorers follow the few basic steps outlined below:

1. Determine the type of floor. This can be done as described above.
2. Determine how far the water has traveled in the floor and how wet the floor is.
3. Determine what layers of the floor are wet and how wet.
4. Extract the floor properly (as described in *Chapter 5 Extraction*), if materials will be saved.
5. Avoid damage to floor surfaces. Repairing floors is expensive and time consuming. Use more aggressive methods first; use disruptive drying only if no progress is being made. Floor mat systems on a negative pressure have been found to be effective on a wide range of flooring systems.

If a wet floor is not drying, it is important to remember the basic rule regarding evaporation: *evaporation will increase as materials are made warmer, as air is made drier, and as air is moved more rapidly across the wet surface.* Under this principle, there are several options for action beginning with the least disruptive:

- Add more direct airflow on the surface and subsurface of the floor.
- Provide drier air to the floor with desiccant dehumidification.
- Dry the floor from both the bottom and top ("sandwich drying").
- Feed air movers that are blowing into the floor with dehumidifiers.
- Warm floor surfaces with heat lamps.
- Remove wet flooring completely.

Each drying challenge is different. The most successful restorers will be ready to employ any and all of these options to reach drying goals in a timely fashion.

Drying Cabinets

When drying cabinets, restorers seek ways to minimize damage while ensuring that the materials dry thoroughly. Specialized equipment is available to generate warm, dry airflow in hidden structural cavities. Most access to a standard cabinet is through the toe kick, a cosmetic cover which hides the under cabinet support. This toe kick is easily replaced in all standard installations. When access is not available through the toe kick, holes are drilled through the shelving in the bottom of the cabinets, through the wall immediately behind the cabinets, or through the back wall of the cabinet itself.

Drying Crawlspaces

Crawlspace drying is a simple process, as long as the number one safety rule of crawlspace drying is followed: *always dry crawlspaces on a negative pressure.* If positive pressure is used to dry a crawlspace, exfiltration will occur, pushing particulate and odors into the living space of the structure. Even the cleanest crawlspace is filled with dust, mold spores, dead insects or worse.

Drying on a neutral pressure is acceptable, but hard to implement in the field. To ensure that a negative pressure is created, more air (CFM/CMH) is removed than is added. Negative pressure, verified with a manometer, is critical to occupant and worker health and safety when contaminants are present.

Other important factors when drying crawlspaces:

- Crawlspaces are dangerous places. Follow all recommendations of *Chapter 2 Safety*.
- Use submersible pumps to remove as much water as possible. Dig a shallow hole in dirt crawlspaces in which to place the pump to pull water to a deeper level (without clogging).
- If using dehumidifiers to dry crawlspace, calculate cubic feet by measuring the outside of the structure and multiplying by average crawlspace height. Then use formulas in *Chapter 8 Drying Equipment*.
- If using air movers, calculate the number needed based on square feet as described in *Chapter 8 Equipment*.
- Heat has fewer limits in unoccupied spaces; use warmer, outdoor air to promote evaporation.

Regardless of the method used, it is important to ensure that the amount of water and the location of water are both identified and monitored. Drying systems are adjusted as progress, or lack of progress, dictates.

Drying Cabinets and Other Interior Cavities

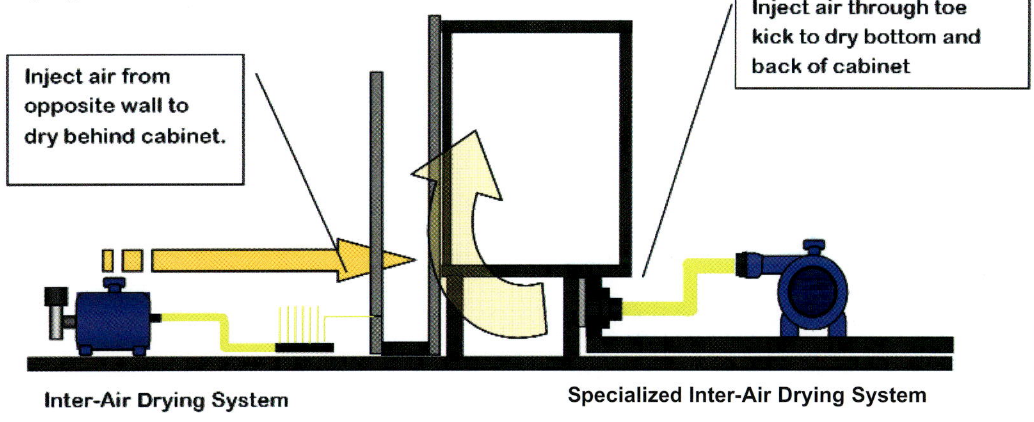

Effectiveness of the drying system is verified in a various ways. For example, if dehumidifiers are being used in a crawlspace, restorers document their performance by using grain depression. If a heat drying system is being used, the negative air output (exhaust) is compared to the outdoor air. The resulting difference in specific humidity is an indication of the moisture that is being removed.

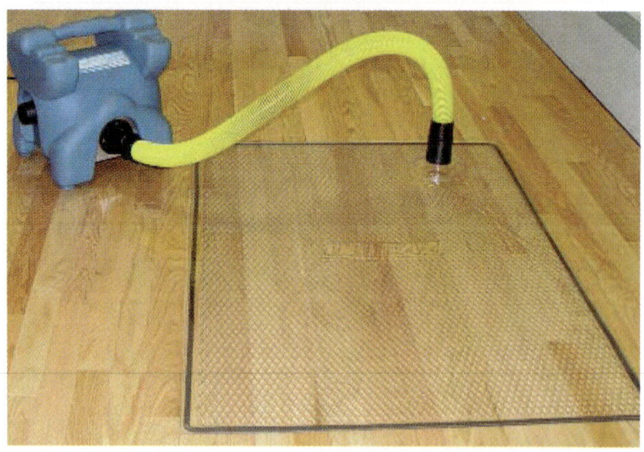

SUMMARY

The most successful restorer will apply the tools and techniques that are best suited to each unique structural drying project. This requires an array of tools to control humidity, temperature and airflow in a variety of circumstances and the knowledge to apply each tool properly. It is critical to diligently and thoroughly evaluate the change in materials over time, and adjust the drying system as progress dictates.

A drying job site is a dynamic environment where change is constant. The change desired in drying is movement toward a pre-loss condition – by first creating a clean, safe environment and then assessing potential damage. Because the job is not complete until materials are clean, dry, and equal or better in appearance and function than when the loss occurred, the impact of each decision must be evaluated closely. Ultimately, decisions must direct the-work toward the most cost effective process available to achieve this end.

THE DRI-EAZ RESTORATIVE DRYING PROCESS

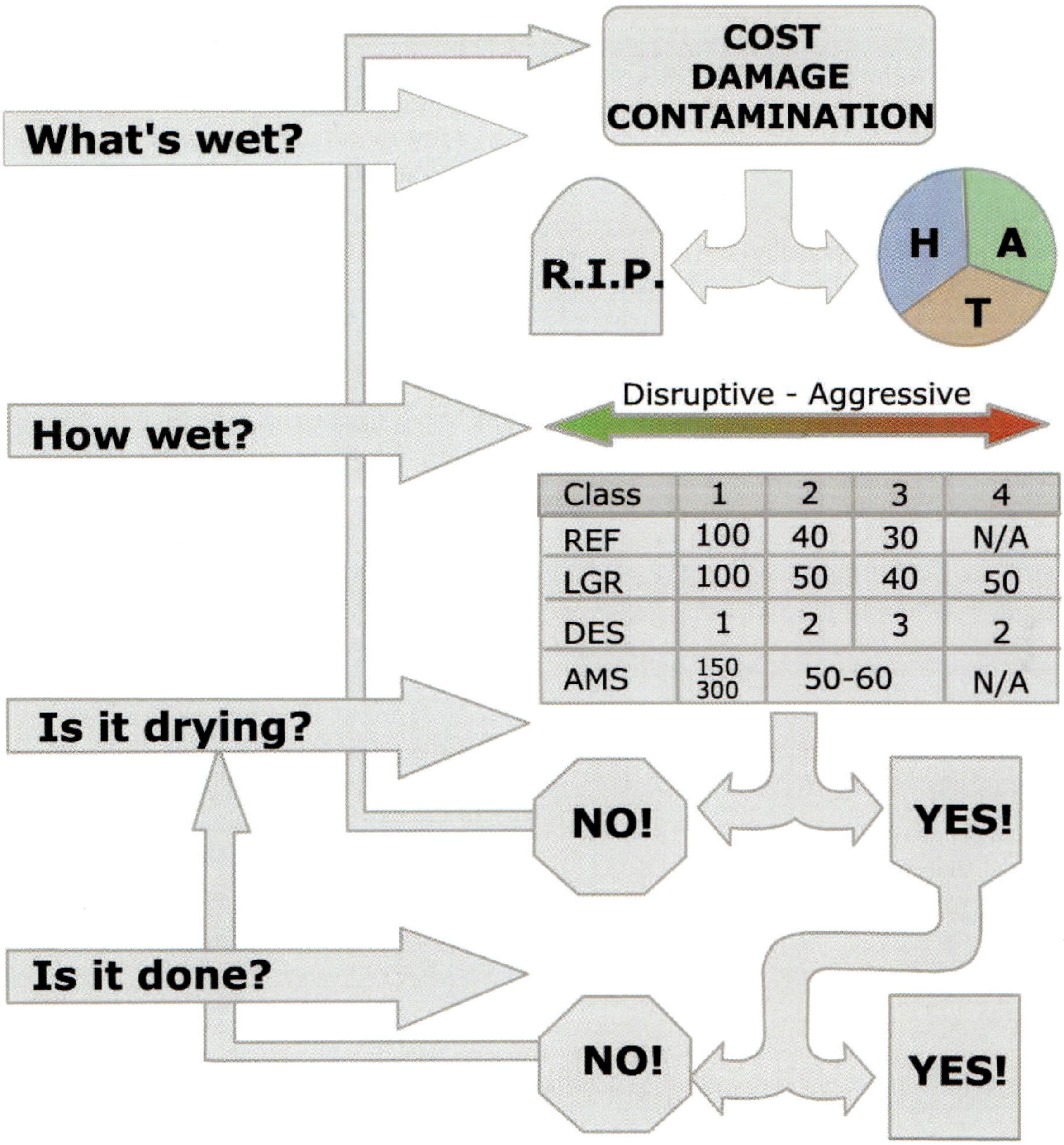

Class	1	2	3	4
REF	100	40	30	N/A
LGR	100	50	40	50
DES	1	2	3	2
AMS	150 300	50-60		N/A

Chapter 10

Microbiology

Typical Fungal Growth

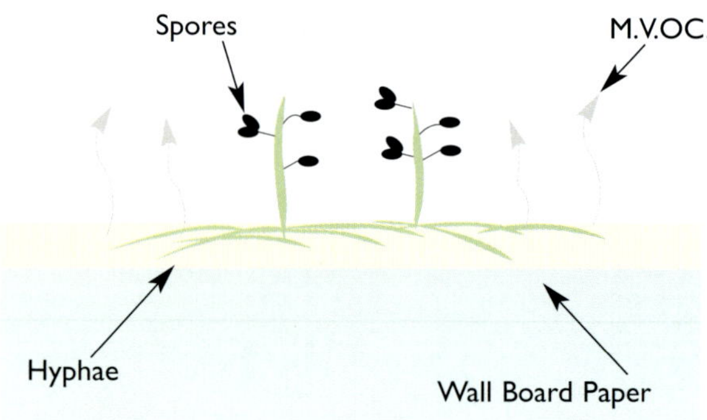

Active microbial growth produces the release of Microbiological Volatile Organic Compounds (MVOCs), and the production of various toxins present in the cellular structure of the organism (e.g., spores, hyphae).

IN THIS SECTION

INTRODUCTION

Restoration of structures that have suffered an abnormal water intrusion serves many purposes, all of which relate to the prevention of secondary damage. The second law of **thermodynamics** dictates that, following a water intrusion, water will begin to *move, evaporate and equalize* throughout the environment. Eventually, the structure would return to normal moisture levels – even if a restorer were never employed. The primary difference between a structure that dries naturally and one dried by a professional restorer is the amount of time necessary for the drying to occur. *The longer a structure remains wet, the greater the potential for secondary damage.*

This secondary damage occurs in many forms including swelling, cracking, splitting, staining, and other cosmetic, aesthetically unappealing changes. One form of secondary damage – microbial growth – can cause structural components to lose their integrity, can potentially impact indoor air quality, and may ultimately result in compromised occupant health.

When structures are allowed to remain wet for extended periods of time, microbial growth begins nature's recycling process. Microorganisms begin to feed on the organic materials in the structure, breaking them down into much simpler organic materials. As this occurs, the levels of **mycotoxins**, **endotoxins**, **exotoxins**, **volatile organic compounds** (odors) and organic debris begin to escalate. Structural components begin to lose integrity and visible damage occurs in the form of staining.

All of this occurs even when clean, relatively sanitary water intrusion occurs – if it is not restored in a timely fashion. In many cases, the water intrusion occurs from a source where many of these contaminants are already present. This water can bring with it various **microorganisms**, **pathogens**, chemicals and other physical hazards. The presence of these contaminants drastically changes the priorities of the restoration contractor. The focus of restoration work always begins with providing a clean and safe environment first, then shifts to drying the structure and contents to prevent further damage.

At times, water intrusion occurs in a structure that contains **pre-existing contamination**. Even if water is relatively **sanitary** as it enters, the contact with organic material and microorganisms as it migrates through poorly maintained carpet and other materials may rapidly degrade its cleanliness.

The condition of the materials may be very unclean for a number of reasons, including lack of proper maintenance and the type of normal use and occupancy. Often restorers can smell the presence of contaminants as soon as water is allowed to dwell in the structure and microorganisms begin to release **microbiological volatile organic compounds** (MVOC).

Knowledge of the microorganisms that will typically be encountered in wet structures is critical in order to properly prescribe restoration procedures. There are various processes and procedures that must be considered when dealing with contamination – all necessary to mitigate immediate and long term health concerns and to prevent secondary damage during the drying process. It is important to understand the effects of incomplete drying, and to be aware of the indicators that microbial growth is occurring, such as the odors that result from incomplete drying.

Water in and of itself does not have an odor. When odors are present, water has contacted a material or organism that can release volatile compounds (odors). Deep, musty or bitter, pungent odors are an indication that the volatile compound is an MVOC, evidence of microbial contamination. Further inspection is necessary in order to find the source. The best way to prevent odors and retard microbial growth in a wet structure is to *ensure that all materials are dried thoroughly and rapidly*. No other procedure is as effective in the prevention of microbial growth.

CATEGORIES OF WATER AND PROCEDURES

A common tool used to evaluate the level of contamination present in a structure that has suffered a water intrusion is categorization of water. The category descriptions in this manual are predominantly derived from the **IICRC** *S500 Water Damage Restoration Standard and Reference Guide, Third Edition, 2006.*

Category 1 water source.

Based upon the source of the water, length of time the water has been allowed to dwell in the structure, the temperature, and pre-existing conditions, it is possible to assess the category and employ appropriate means of restoration.

The category of water will influence what materials will be saved, and which will be replaced. It will also influence what type of safety equipment that will be used, and who will be allowed access to the structure during the restoration process, especially before and during the cleaning phase. Properly categorizing water is a critical component in successful restoration.

A structure that has been merely dried to a pre-loss condition has not necessarily been restored. A structure that has been dried – but where cleanliness has been ignored – can lead to an escalation of contamination, spreading contaminants to unaffected areas.

Restorers will diligently investigate structures for indications of contamination, assigning a category only after a thorough investigation.

Category 1 - Clean Water

Water with no significant risk of causing sickness or discomfort is said to be **Category 1** water, or "clean." Generally, Category 1 water intrusions are the result of compromised sanitary water supply lines. In order for a situation to remain a Category 1, water must not have sat in a structure for more than 72 hours, and materials affected must be clean and well maintained. If odors are present in the structure, further investigation is necessary – until the source of the odor is found. Odors indicate that the cleanliness of the water is not Category 1.

Microbiology

When Category 1 has been established, structural drying can continue with no substantial change (to standard procedures) for addressing contamination. It is possible to effectively restore the structure to a pre-loss condition by thoroughly and rapidly drying materials, and by replacing only materials with permanent structural or aesthetic primary damage.

Over time, the category of water can degrade. Even in a clean environment, many organic materials and a number of **microorganisms** will be present – and the addition of moisture allows for these organisms to feed on organic materials and multiply. The faster a material is dried, the less this growth is allowed.

For this reason, restorers employ procedures to mitigate this growth when responding to a Category 1 water intrusion, such as cleaning the affected carpets using a hot water extraction method prior to beginning the drying process. This reduces the organic load present in the carpeting, thereby **mitigating** the growth of microorganisms. Restorers also use **antimicrobial** agents designed to reduce the number of organisms present and retard the growth of additional microorganisms while the structure is drying.

Various products are available to help inactivate a number of microorganisms, leaving them unable to reproduce (making them non-viable). Each antimicrobial product comes with its own set of specific instructions and guidelines – and each must always be used in 100% accordance with its label.

Although the majority of a structure may be categorized as clean, it is important to evaluate each area affected separately, and to consider the structure's cleanliness *before* and *after* drying is complete.
One area of particular significance is the building's HVAC system. Building HVAC systems often contain a high degree of soils and dusts that have built up over time in the mechanical components, in duct work, in condensate trays and in the filters. The addition of moisture can lead to a more rapid escalation of contamination because of the abundance of food and, in some cases, pre-existing contamination.

Before using an HVAC system that has been affected by a water intrusion, it must be examined closely. Even after drying has been completed, the system must be evaluated to determine the need for any additional service.

Aquarium water is an example of Category 2.

Category 2

Water that does carry a significant degree of chemical, biological and/or physical contamination is said to be **Category 2**. Water sources that can result in Category 2 damage include: aquariums, waterbed leaks, toilet bowl overflows that originate from the sanitary water supply yet contain urine from the bowl, dishwasher discharge, clothes washer discharge, and water that enters the structure from **hydrostatic pressure** (from below grade).

In order to remain a Category 2, water must not be allowed to dwell in the structure for an extended period of time. If the water source was originally a Category 1, the water will degrade to a Category 2 after approximately 72 hours. When water is a Category 2, it will continue to degrade and become a Category 3 after 48 hours.

The amount of time it takes for water quality to degrade depends upon two primary factors: the temperature of the water and its content. The times expressed in this chapter are referring to normal indoor temperatures, and are very general. The water's content will also dramatically impact the rate at which its quality degrades. The more contaminated the water, the faster it degrades.

When structures are affected by Category 2 water, special steps and procedures are necessary in order to return the structure to a pre-loss condition. Cleaning procedures *must* be employed before the drying process can continue. At a minimum, affected carpet underlay (pad, cushion) must be removed and disposed of, and carpet must be thoroughly cleaned using a hot water extraction method.

Restorers then employ appropriate antimicrobials to mitigate growth of microorganisms, especially if any porous material is to be cleaned and restored. As with any water intrusion, once the affected materials have been completely cleaned, thorough and rapid drying is necessary to prevent further damage.

Category 3

When water intrusion results from a grossly unsanitary source, carries pathogenic (disease causing) agents, or when water has dwelled in a structure for more than 120 hours, it is said to be **Category 3**. Examples of Category 3 water sources include: discharge from toilets that originate from beyond the toilet trap (from the sewer or septic system), and intrusions from the surface of the ground into the structure (flood waters).

Category 3 water intrusion – water originated from the sanitary sewer system.

Regardless of the category of water from its initial source, water can be deemed a Category 3 if it is allowed to dwell in the structure for an extended period of time. Typically, this period of time is approximately 120 hours, or five days. The amount of time necessary depends upon temperature and the water's content after migration through the structure.

Many procedures are necessary to address cleanliness and safety when dealing with Category 3 water intrusions.

- Several forms of PPE are used, including rubber gloves, respiratory protection, body coverings, rubber boots and splash goggles.
- Worker and occupant health and safety are the first priority on every project. Areas affected by the water intrusion are marked and posted as hazardous.
- Individuals with compromised immune systems are evacuated from the structure until it has been judged safe for occupancy, e.g. the very young, the elderly, those who have undergone recent surgery or chemotherapy, and those whose immune systems are suppressed by conditions such as AIDS.
- Various cleaning and decontamination procedures must be used, including the removal of all highly porous materials (e.g., carpet, underlay [pad, cushion], wall board, insulation) affected by the water intrusion.
- Surfaces that are to be restored must be thoroughly cleaned.

Very few exceptions exist to these rules when dealing with Category 3 water. If attempting to restore any highly porous item affected, restorers must be sure to properly apply a **biocide** (immersion); allow proper dwell time; perform intensive cleaning; and employ third party testing. This process requires a great deal of expense and will not therefore be practical for most highly porous materials. One example of an exception: a valuable oriental rug, where thousands of dollars in value could support repeated applications of this process.

Additionally, restorers apply an appropriate biocide during and after the demolition process to help control microorganisms and increase cleaning effectiveness. They must be familiar with the biocide's product label and local laws and regulations that govern its use.

After an initial assessment and the determination of necessary procedures, it is important to ensure that all materially interested parties are made aware of the work that will be done. If resistance is encountered, restorers should ensure that the reasons for the procedures have been communicated.

TIPS

It is critical to thoroughly investigate the water's source, path of migration and potential level of contamination in order to successfully restore the structure to a condition that is safe for re-occupancy – and to ensure the protection of worker and occupant health during restoration. The categories discussed above, and the practices suggested, are important to consider.

When resistance is still encountered and performing services properly becomes a challenge, restorers stop work immediately to prevent liability. No amount of documentation, waivers or releases will offer complete protection if the work is done in a way that does not diligently provide for occupant safety during and after the restoration process.

MICROORGANISMS

The world is populated by many living things of varying size, shape and function. However, the greatest population on Earth – both in total weight and number – is the microorganism.

Microorganisms are living things that are too small to see with the naked eye. Microorganisms include **fungi**, **bacteria**, **protozoa**, **parasites** and **algae**. Although they are not living, **viruses** are often considered microorganisms as well.

Each of these groups of microorganisms can be found throughout the many environments on Earth. In fact, many are encountered in the normal indoor living environment. Because they are present virtually everywhere, their existence is of no significant risk in and of itself. However, the number of organisms present – combined with the strength of the individual immune system, pathways for potential exposure, and individual sensitivities – define the level of risk.

Structures that have been exposed to water over a long period of time are likely to see an amplification of microorganisms. Predominantly, when water is allowed to dwell in and near organic materials, fungal growth will be the most apparent microbial amplification.

When water intrusions occur from unsanitary sources, the primary concern is not with fungi, but with the presence of bacteria, viruses and parasites. When unsanitary water contains high degrees of organic material, or migrates across or through soiled structural components, bacteria especially can amplify at phenomenal rates.

HIV Virus structure

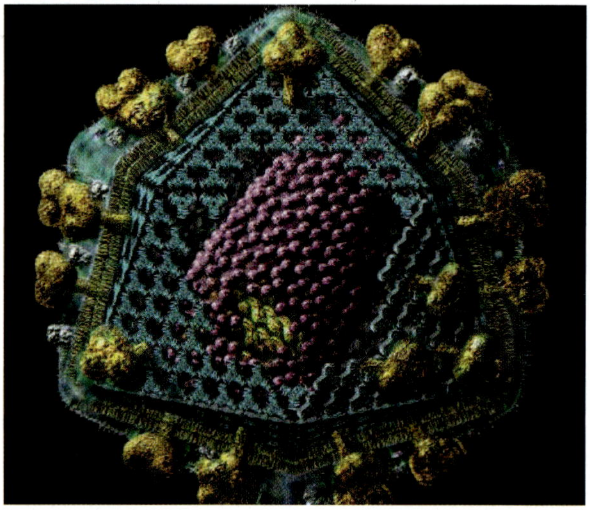

Viruses vary from bacteria and mold in that they are not living organisms.

Although microorganisms are a critical part of nature's recycling process, and are critical in such processes as the manufacture of medicine, foods and other consumable goods, they can be very damaging. The microorganism amplification can present a potential risk to both structural integrity and human health. Fungal amplification runs the greatest risk of compromising structural materials. Bacteria, viruses and parasites each present a significant risk to human health, responsible for such diseases as salmonella, hepatitis, HIV and tetanus.

Because the risks are high, it is critical to:

- Properly remediate contaminated structures
- Use proper procedures to prevent an amplification during the restoration process.

There are some important modifications to standard restoration practices that must take place when dealing with the two primary types of microorganisms found in water damaged structures and materials: *fungi* and *bacteria*.

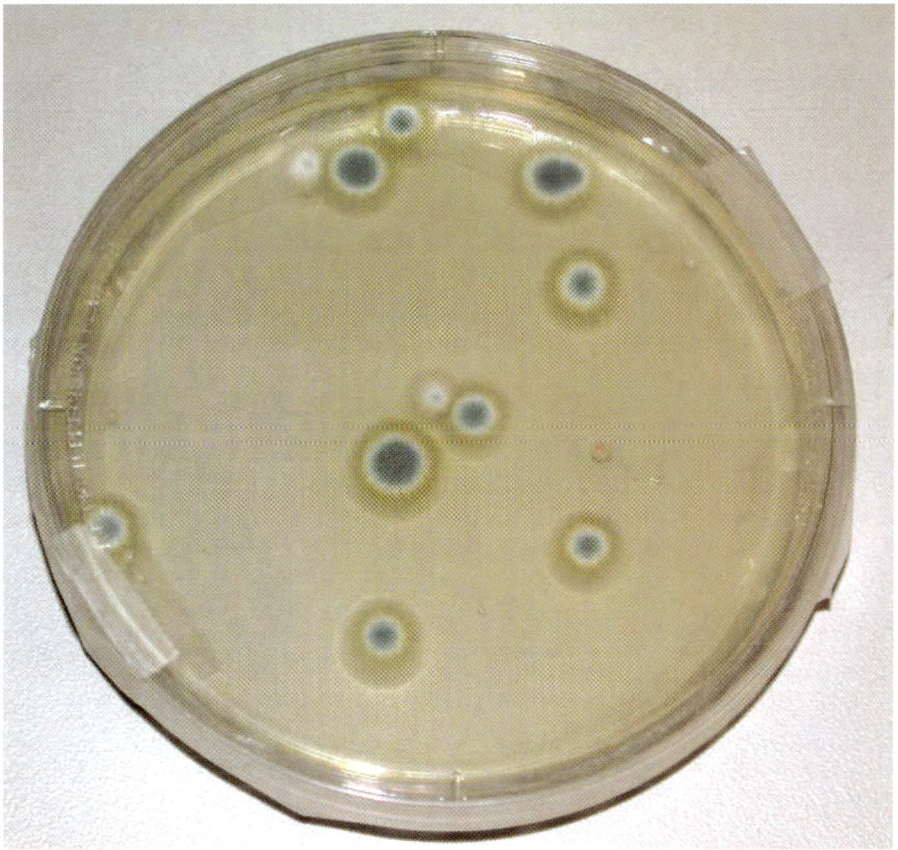

Fungal growth in a laboratory petri dish

Fungi

The microorganism most apparent in structures that have had prolonged contact with relatively sanitary water is **fungus**. Fungi are the natural recycling plants in nature, breaking down an array of organic materials. Without fungi, dead trees, leaves, branches, grasses and other organic materials would pile up over time and never completely decay. Fungi are also responsible for the production of such products as beer, cheese, bread and penicillin.

There are five types of fungi: *mold, mildew, mushrooms, slimes* and *rusts*. When structures are allowed to remain wet, the fungus of concern is **mold**. Mold can germinate within wood and begin to break down the material, attacking the cellular pulp and destroying its structural integrity. This type of fungi is commonly referred to as *rot*, but is actually a member of the mold family (e.g., **"dry" rot** – merulius lacrymans; **"wet" rot** – coniophora cerebella). Rot can occur in wood that is visibly wet, or above fiber saturation point. When this occurs, the rot is referred to as a wet rot. Rot can also occur when wood is visibly dry, below its fiber saturation point, yet wetter than 20% moisture content. When this occurs, it is referred to as dry rot.

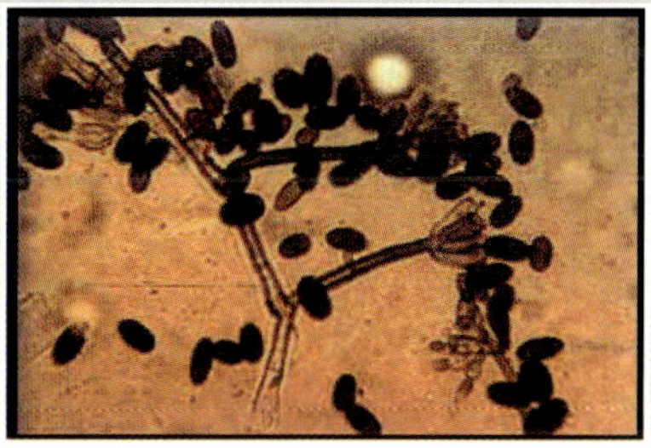

Stachybotrys under a microscope

When moisture levels are sufficient (16% moisture content for wood), mold can begin to germinate and grow on the surface of virtually any material. Surface molds can feed on the dusts and soils present across every interior surface, in carpets, etc. They can also grow on cellulose materials like paper and wood. Surface mold is similar in appearance to the traditional plant; it is comprised of **hyphae** that are very similar to roots, which form **mycelia** – a large complex mat of roots. The mycelia act as a vapor retardant, trapping moisture at the surface of a material to support continued growth. **Conidiaphores**, similar to a trunk on a small tree, support the growth of **conidia**, **sporangiums** and **spores**.

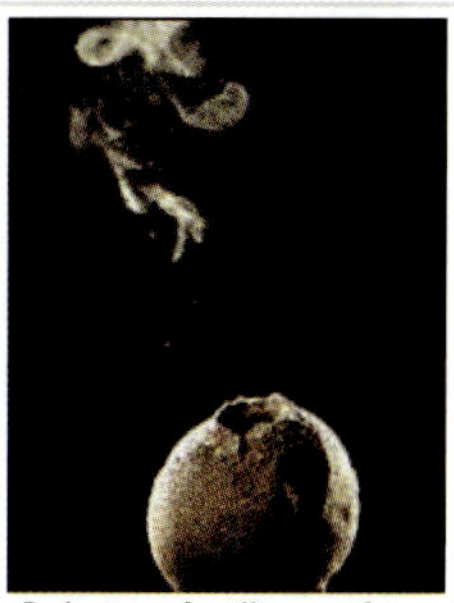

Release of millions of spores

Life begins when a mature spore, released some time earlier from a

FUNGI AND WOOD MOISTURE CONTENT

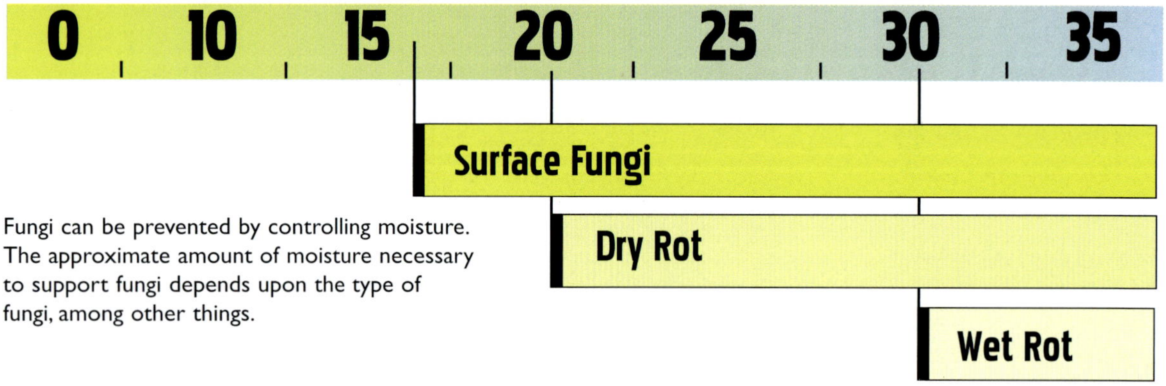

Fungi can be prevented by controlling moisture. The approximate amount of moisture necessary to support fungi depends upon the type of fungi, among other things.

sporangium (mold structure smaller than a pin head containing thousands to millions of spores), settles on a surface with ample moisture. The cell begins to absorb the moisture, causing it to swell. Most spores, as they swell, begin to release a mixture of the absorbed moisture and an enzyme contained either on the spore wall or within the spore itself. Once this enzyme solution contacts an organic material, the reaction produces the substances necessary for the spore to continue its **germination.**

As a spore becomes an actual "colony" of mold, several significant processes are occurring. First, the process of digestion occurs externally as mold breaks down organic material. This creates a by-product,

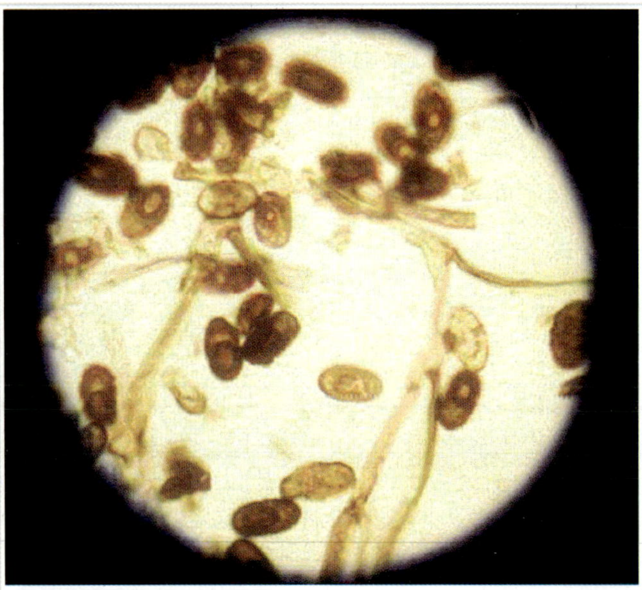

Stachybotrys spores, conidia and hyphae fragments

known as a ***Microbiological Volatile Organic Compound*** *(MVOC).* MVOCs, being volatile, are released into the surrounding air. They are responsible for the odor associated with mold growth, and are typically an earthy, musty odor. When these odors are present, they indicate that active microbial growth is occurring.

This means that both elevated moisture and mold are present. Although MVOCs can be prevalent in an area with active mold growth, they are not directly responsible for any known human ailment.

Second, as mold grows, it can produce a toxic chemical substance. The toxin produced depends upon the type of mold, but all mold related toxins are referred to as **mycotoxins**. Several mold species are known to produce mycotoxins, including: A. fumigatus, A. versicolor, Trichoderma, Exophiala, Stachybotrys, Phialophora, Fusarium, Ulocladium and Penicillium.

Mycotoxins give mold a competitive advantage over neighboring microorganisms competing for food. They are most commonly released when mold cells are stressed, and are typically carried within fragments of the hyphae and in spores. Unlike MVOCs, some mycotoxins have been shown to be toxic to humans. Human exposure to mycotoxins can be increased when airflow is allowed to aerosolize particulate from active or inactive mold growth, when

Special procedures and personal protection equipment are necessary when microbial growth is present.

environmental conditions are changed, and when mold growth is allowed to come into direct contact with skin.

With this in mind, it is important to consider the impact of drying equipment when mold is present. As with any contaminant, mold is removed or contained before any environmental controls are manipulated to support evaporation. **Remediation** of microbial growth is completed in a manner that is highly sensitive to the spread of spores and hyphae fragments. The use of antimicrobials and biocides is for the management of growth only, and not for elimination of the contaminant. Remediation must be followed by proper removal of the mold growth.

Removing mold growth often necessitates the removal of the affected material, especially when materials are highly **porous**. Semi-porous materials can often be cleaned aggressively, even sanded or media blasted, to remove the surface growth completely.

Note: Workers in areas with extensive mold growth must wear appropriate respiratory, eye and hand protection to prevent exposure to mycotoxins.

Bacteria

When structures are affected by unsanitary water from outdoor flooding, sewage systems and septic systems, bacteria are present in very high numbers. The concerns related to bacteria are similar to those regarding fungi. Bacteria grow and feed on organic material, and they produce MVOCs and toxins. Under favorable conditions, the number of bacteria can double every 15 minutes.

Odors from bacterial amplification are typically much more "bitter" and pungent than odors associated with mold. **Toxins** created by bacteria are referred to as **endotoxins** and **exotoxins**; the difference being whether they reside inside or outside the bacterium's cell wall.

Bacteria are also similar to mold in that they exist nearly everywhere in nature. They are predominantly present in soil, but can be found in common

Microbiology

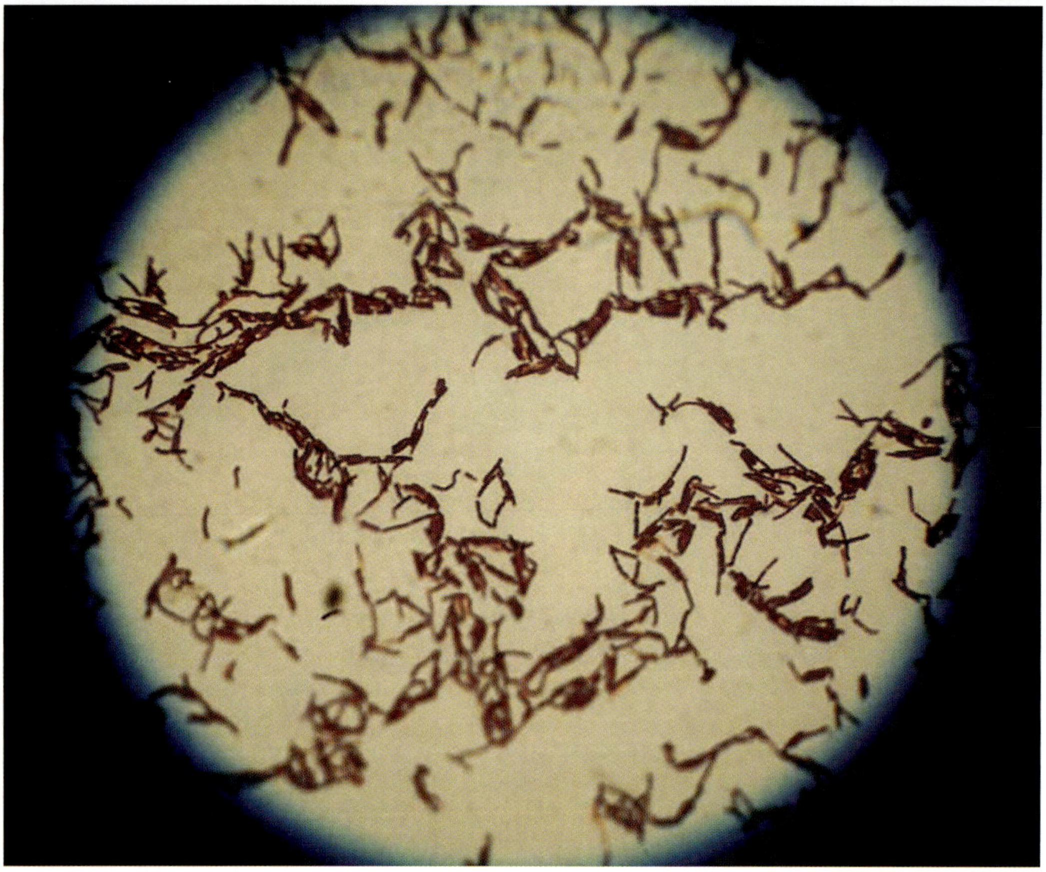

Bacterial amplification under a microscope

household dust and dirt, on human skin, and on many surfaces where bio-film (organic residue) exists. They are relatively small (.4-10 **microns**), single-celled microorganisms.

A significant difference between bacteria and mold is that there are many more common diseases known to be caused by bacteria. One such ailment is Legionnaire's Disease. Bacteria reproduce much more rapidly than other microorganisms and can quickly threaten the indoor environmental quality.

ANTIMICROBIALS AND BIOCIDES

Various products are used to control, regulate or eliminate microorganisms throughout the world. They are used in engineered building materials, home cleaning products, industrial cleaning products, personal hygiene products and other related goods. Extensive studies, research and testing have examined their safety, effectiveness and use.

No tools in the restorer's arsenal are as extensively tested, or regulated, as antimicrobials and biocides. Before deciding to use an antimicrobial or biocide, it is critical to first understand the terminology, technology and chemistry behind the individual product.

- **Antimicrobial** – This term has two common definitions: (1) a product used to control, inhibit or eradicate microorganisms; and (2) a product applied during the manufacture or production of a material to add resistance to microbial growth.

- **Biocide** – A product intended to destroy or leave non-viable one or more types of microorganisms.

Every product used, in other words, is an antimicrobial. Biocides are simply a type of antimicrobial that has been designed to kill or destroy. There are three

levels of biocidal activity, according to the **Environmental Protection Agency (EPA)**:
www.EPA.gov

- **Sanitizer** – A cleaning or disinfecting treatment designed to reduce the number of pathogenic microorganisms to a safe level. This is the lowest level of biocidal activity.
- **Disinfectant** – A solution designed to destroy or inhibit the growth of microorganisms, but not necessarily their spores.
- **Sterilizer** – A product designed to destroy all microorganisms (fungi, bacteria, viruses, etc.) and their spores.

Several terms can be used to specify the type of target organism a product or procedure has been designed to control or destroy. These terms each begin with the target organism, and end with one of two suffixes:

- **"-stat"** – Terms used to describe an antimicrobial or biocide that end with this suffix are intended to control or regulate growth of the targeted organism (e.g., a fungistat controls the growth of fungus).
- **"-cide"** – Terms used to describe an antimicrobial or biocide that end with this suffix are intended to destroy the target organism (e.g., a virucide would destroy viruses).

Active Ingredients

Each **active ingredient** commonly found in antimicrobials and biocides used in restorative drying has its own strengths, depending upon the intended application site, target organism and environmental conditions.

Choosing an antimicrobial begins with understanding the efficiencies and deficiencies of each of these active ingredients.

Alcohol – Isopropanol, or "rubbing alcohol," is a very effective antimicrobial. Unfortunately, it is extremely volatile and highly flammable. This leaves for very little dwell time and creates a tremendous safety hazard.

Aldehydes – Commonly found in the form of formaldehyde or glutaraldehyde. Aldehydes are

EPA DEFINITIONS – LEVELS OF BIOCIDAL ACTIVITY

SANITIZER	**LOWEST** Reduces the number of microorganisms to a safe level, generally labeled as "stats" that control or regulate growth.
DISINFECTANT	**99.9997% EFFECTIVE** Products that free the application site of infection, killing or destroying at least 99% of the target organism. Generally labeled as a "cide."
STERILIZER	**HIGHEST - 100% EFFECTIVE** Product that has been proven to eradicate all target organisms from a surface, including the spores.

Three levels of biocidal activity defined by the EPA

non-ionic, allowing them to be used on stain resistant carpets without removing the stain-resist qualities. They are also highly inactivated by organic materials.

Hypochlorites – Effective against a broad spectrum as a disinfectant and sanitizer, with good deodorizing properties. Hypochlorite is inexpensive and readily available. Unfortunately, it is a strong bleaching agent, corrodes metals, is inactivated by organic matter and has limited cleaning (soil suspension) ability.

Hydrogen Peroxide – Works well as a "stat" against many organisms, but is effective for a relatively short period of time. Hydrogen peroxide is very sensitive to organic materials, losing effectiveness rapidly. It can bleach fabrics and cause corrosion.

Phenolic Compounds – Commonly used for sewage, phenols are very effective against a wide variety of microorganisms. Phenols are more impervious to organic matter than other antimicrobials. Phenols are easily absorbed through the skin (dermally) and require a high degree of caution when used.

Quaternary Ammonium Compounds – Commonly referred to as "quats," they are the common active ingredient in both institutional and consumer products. Quats are very safe relative to other active ingredients when properly diluted, and effective against a wide variety of organisms. Quats have good cleaning (soil suspension) properties with a high alkalinity.

Ozone – Very effective for destroying volatile organic vapors and eliminating odors. Difficult to support sufficient penetration into many materials and can be dangerous in high concentrations. Reacts with background chemicals forming new compounds.

Antimicrobials and deodorizers are important tools used in the successful restoration of many water damaged structures.

Antimicrobial

Deodorizer

Using the Right Antimicrobials

Choice of the appropriate antimicrobial requires examination of the following factors:

- The type and extent of contamination
- Potential adverse health effects
- Other procedures available to prevent or control microorganism growth

Many products contain one or more antimicrobials, even common household products.

In conditions with high risk of exposure to organisms that can cause sickness, discomfort or disease, restorers will choose to use a registered biocide. This is especially true in circumstances where the building occupants are at a higher risk because of a compromised immune system.

Where cleaning and drying procedures alone eliminate any risks, it is important to use procedures that, in and of themselves, do not create *additional* risk. The risk associated with a chemical agent, in other words, is weighed against the risk presented by microorganisms. *If the chemical agent carries a greater risk, then it is not used.*

Many antimicrobial products can act as **sanitizers**, deodorizers and cleaners across a broad spectrum of building materials, while they are not necessarily bio-

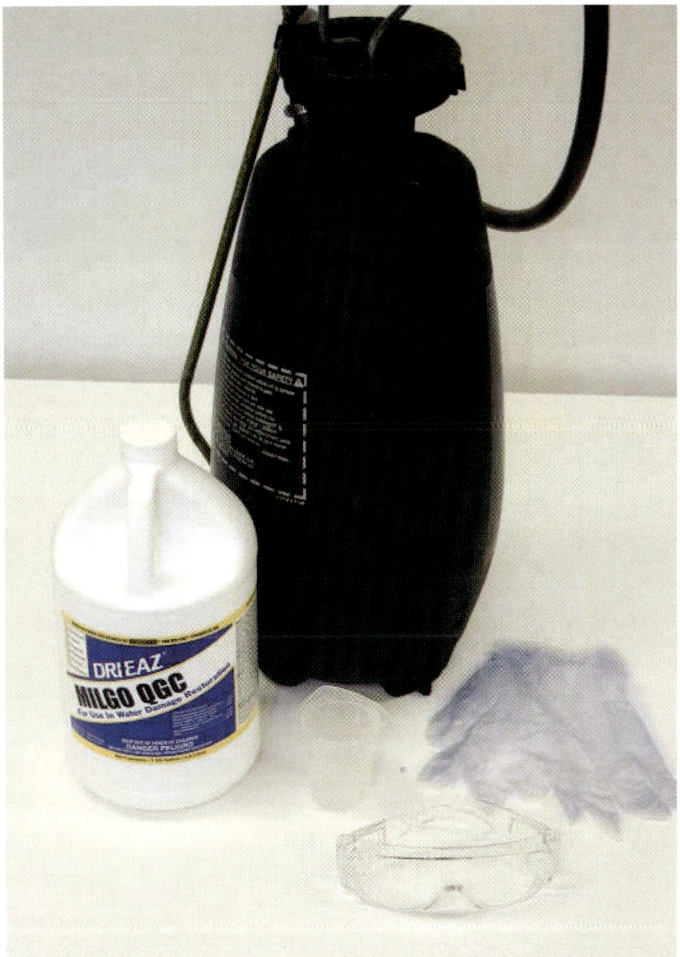

Using antimicrobials properly requires measurement and personal protection.

cidal on all surfaces. Products that do work as a biocide across a broad spectrum of materials for an extended period of time are generally those products with greater safety risks and, in some cases, greater risk of damaging the materials themselves.

To understand what materials the product will treat effectively and what health and safety risks they present, it is important to become very familiar with the product's label. The use of a biocide may or may not be advisable. However, general procedures that are antimicrobial in nature are *always employed*. The art is matching the right procedure to the job, just as with any other restoration practice.

Using Antimicrobials Safely

Whenever a chemical antimicrobial or biocide is used, restorers must follow the label directions explicitly. Any product that is labeled as an antimicrobial will have an EPA registration number and an approved EPA label. It is a violation of federal law to use a product in a way that is inconsistent with its labeling. On an EPA registered label, several vital pieces of information can be found, including:

- **Approved Application Sites** – The product will have been tested on and approved for limited application. Many antimicrobials are not listed as "approved for application to carpet." Because carpeted surfaces are a common surface found in homes, always be prepared with a product that is approved for treating them.

- **Application Methods** – Many methods have been used to apply antimicrobials. The most common method approved for the products available in our industry is through a low pressure spray nozzle. Many products are *not* approved for "fogging," or application through an ultra low volume **(ULV) droplet fogger**.

- **Target Organisms** – The label will provide information on the organisms against which the product has been tested. For example, when

The Product Label

Technicians should read the entire product label and follow instructions exactly.

using an antimicrobial for sewage, it is important that the common bacteria and viruses *associated with sewage* are listed (e.g. E. Coli, Pseudomoras, Hepatitis).

- **Application Rate/Dilution** – Antimicrobials are most effective at their proper dilution. In order for the chemical to perform its work, there must be enough water present. It is the solution that acts as an antimicrobial, not the chemical alone. Proper dilution is critical.

- **Dwell Time** – This value refers to the time necessary for the solution to remain in contact with the targeted organism in order to be effective.

- **Personal Protection Equipment** – The label will provide information on what PPE must be worn during the application.

- **Evacuation** – Information regarding structure vacancy will be provided, including the amount of time the structure must be vacated after application before re-entry.

- **Disposal** – Instructions will be included for what should be done with the empty container, unused concentrate and unused solution.

- **Storage** – Certain requirements for storage are typically necessary for maintaining chemical effectiveness and preventing potential hazards.

This list represents only a portion of what is available on the product label. It is important to be familiar with each component of the label and follow these directions explicitly. Failure to do so may result in a fine, lawsuit or even imprisonment.

When using a **biocide**, it is important to first conduct a customer briefing with the building owner and occupants. Make the customer aware of the product that will be used, why it is being used, and how long they must vacate the structure. Most antimicrobials require a vacancy of the structure before, during and after application. Also, it is important to obtain a signature on an informed consent form, documenting this procedure and the customer's approval. Finally, building owners and occupants are provided a **Material Safety Data Sheet (MSDS)** upon their request.

PRESSURIZATION

When contaminants are present in a structure, consider the use of **pressurization** to control the spread of airborne contaminants. Pressures are a great ally in controlling airborne contaminants when used properly. However, they lead to disaster if *not* used properly. Because the result of pressure differences is so dramatically advantageous or disastrous, pressures are closely monitored in a contaminated environment. The two most commonly used tools for monitoring pressures are **manometers** and smoke generating devices (e.g., smoke pencils).

First, restorers formulate a pressurization plan that will best suit the job; then, monitor the pressure and the resulting air movement to ensure proper containment. The most critical rule that applies to air pressures is to keep the most contaminated area negatively pressurized, and the cleanest area positively pressurized.

Pressures are relative, in that one pressure level can be negative and positive at the same time. Negative pressures are created by removing air from a space, while positive pressures are created by adding air to a space. When air is removed from one room in a structure, in other words, it becomes negatively pressurized. However, it is only negatively pressurized when compared to neighboring rooms. The room –

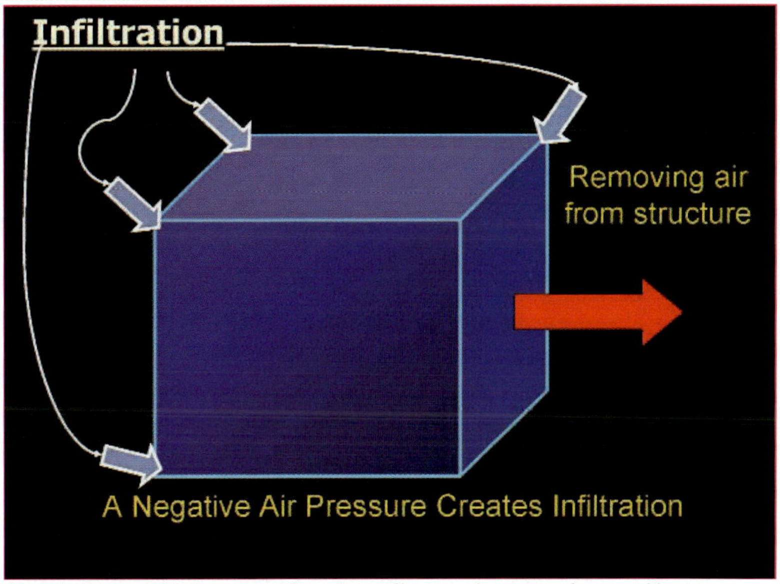

Negatively pressurizing a room will result in infiltration, drawing air from neighboring spaces.

even when under a negative pressure – is still quite positive when compared to, say, outer areas.

This comparison can be made several times over, throughout the structure. The room being negatively pressurized will draw air from the surrounding air spaces. It will draw air from, for example, the hallway. As air is drawn from the hallway, it too becomes negatively pressurized, yet remains positively pressurized compared to the initial room discussed. It is easy to see how pressures can become complicated, as air moves from room to room and from space to space – seeking equilibrium in air pressure.

As soon as an inequality is created, nature seeks to make things equal again. As long as the inequality remains, nature will continue to seek balance, or equilibrium. This will result in flow of air between each area of pressure difference.

Negative Pressurization Causes Infiltration

Negative air pressures are best used in areas that contain the highest level of contamination – whether the contaminant is microbial, chemical, physical or even just water vapor. Negative air pressure is created by removing any volume of air from a space. Where a negative pressure is created, air will move from neighboring spaces to "make up" for the air removed. The air that does enter the space will enter through cracks and crevices – and will carry heat, humidity, particulate or any other airborne agent. This process is referred to as infiltration.

When employing negative air pressures, it is possible to influence the infiltrating air, but not to entirely control it. The air exhausted to create the negative pressure, however, can be *completely* controlled. This air can be filtered, dehumidified, conditioned or even deodorized before it is delivered to the next air space. (Normally it is simply blown outdoors.)

The desired result of the negative air pressure is the fact that all air is entering, or infiltrating into the space that is negatively pressurized. This prevents any air in this space from exiting into unaffected areas, and allows for a great deal of control over any particulate or vapor in that air.

Caution must be taken when using negative air pressure. Negative air pressures will also draw air from all neighboring air spaces. Restorers closely evaluate each neighboring space and consider the temperature, humidity and cleanliness of the air. Should it be drawn into the structure?

Strong negative air pressure can cause combustion appliance exhaust to back draft into the structure. This can result in a build-up of **carbon monoxide** in the structure, causing asphyxiation and even death.

> Two examples of when a negative pressure will be appropriate are:
>
> **1)** *when drying crawlspaces* or
> **2)** *when drying wall cavities with suspected contamination.*

Crawlspaces are commonly unclean, unsanitary environments. When dried, they are held under a negative air pressure to prevent volatile organic compounds and particulate from entering the air space above.

Wall cavities that might to be contaminated are dried much the same way. Inter-air drying systems are used under negative air pressure, causing any exchange of air to infiltrate into the cavity, thereby preventing any potential contaminant from spreading to unaffected areas. Air pulled from the cavity by the inter-air system is blown outside or filtered using an air scrubber with HEPA filtration.

Positive Pressure Causes Exfiltration

Positive air pressures are created by adding air to a space. Whenever air is added to a space, an inequality is created in air pressures. This inequality will result in the **exfiltration** of air into neighboring air spaces. Air will exit through cracks and crevices, and will carry with it any airborne particulate or vapor.

Positive pressure is used whenever a space must remain uncontaminated or dry compared to neighboring areas. However, creating a positive pressure is often quite challenging.

In order to create a positively pressurized clean room, access to an unlimited supply of clean air is critical. In order to create a positively pressurized dry room, access to unlimited dry air is needed. This air is typically only available when the outdoor air is usable as the clean or dry air. Positive pressures are only used when ex-filtrating air will not adversely affect unaffected areas. An example of when positive pressures are beneficial is when ex-filtration will occur through targeted materials that need to be dried, forcing warm dry air through affected interstitial cavities.

It is important to ensure that positive pressures are *never* used in a space that contains a known or suspected contaminant. Ex-filtrating air could carry the contaminant to unaffected areas. This is true regardless of the contaminant type, whether it is a particulate, volatile organic compound, chemical or simply water vapor.

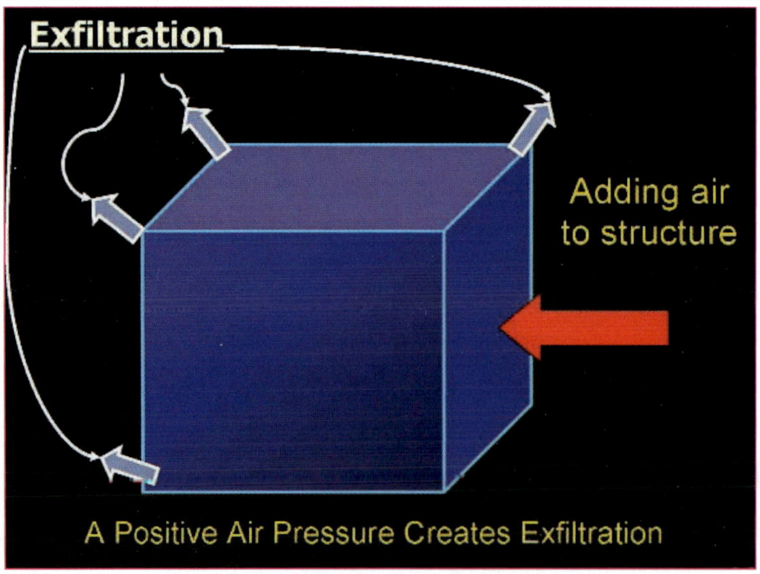

Adding air to a space will result in exfiltration to neighboring spaces.

THIRD PARTIES

In cases involving extensive contamination, it is necessary to involve an independent, third party evaluator or other professional to assess structural damage, microbial problems and other potential health and safety risks. Restorers use a third party in situations that involve severe public health issues, high risk occupants living or working in the structure, or extensive microbial contamination that can affect worker or occupant health.

There are a number of third party experts and specialists to consider, including **industrial hygienists**, physicians, microbiologists and in some cases attorneys. The simple rule as it applies to a third party is: consider expert advice and guidance when in doubt, or when concerned about the amount of risk and liability associated with the impact of restoration related decisions.

In some cases, the building owner can be advised to consult an independent third party, removed from the restorer and the insurer. Cases that require the intervention of an independent third party are those that involve conflict. More information on this subject is available in *Chapter 4 Inspection, Monitoring and Documentation*.

Appendix

TERMS AND CONDITIONS FOR USE OF FORMS:

Dri-Eaz Products, Inc. provides complimentary sample materials as a courtesy to our customers. However, Dri-Eaz does not give legal advice. Before using any legal document, you should consult an experienced lawyer who is familiar with the water damage restoration industry and the challenges you face.

The law of each jurisdiction is different. Some states have cancellation notice requirements, finance charge restrictions, font size requirements and a variety of other rules. These forms do not contain state-specific requirements, and are not intended to be legal advice.

These materials have been collected from different sources and from different regions, each of which has unique characteristics. These forms are not intended to be comprehensive, but will hopefully help Dri-Eaz customers identify issues to discuss with their attorneys.

By using portions of these forms, you agree to release, indemnify and hold harmless Dri-Eaz Products, Inc. and its successors and employees from any liability, claims or damages that result from the use of these complimentary sample materials. Your permission to use these materials is conditional upon your agreement to the foregoing.

Call Report Form

Job Information

Date: ____________________ Time Called In: _____________ Date of Loss: ______________ Job No: ____________

Contact: ____ Relationship: ___________________________________

Job Address: ____ Site Phone: ____________________________________

City: ____ Work Phone Mr: _________________________________

State/Prov: _______________ Zip/PC: __________________ Work Phone Mrs: ________________________________

Other: ____

Insured Information

Insured: ____

Bill Address: _______________________________________ Home Phone: ____________________________________

City: ____ Work Phone Mr: _________________________________

State/Prov: ________________ Zip/PC: _________________ Work Phone Mrs: ________________________________

Motel: ____ Phone: ___________________ Ext: _______________

Other: ____

Insurance Information

Ins. Company: _______________________________________ Policy : ______________________________________

Ins. Agency: ____ Contact: ______________________________________

Agency Address: _____________________________________ Phone: ____

City: ____ State/Prov: ________________ Zip/PC: ____________

Adjusting Co: _______________________________________ Adjuster: _____________________________________

Adjuster Address: ___________________________________ Phone: ____

City: ____ State/Prov: ________________ Zip/PC: ____________

Directions to Job	
Map Page:	Coordinates:

Telephone Questions:

Is there any standing water?	Yes No	Is there electric power available?	Yes No
Has the source of water stopped?	Yes No	Is the heat system still working?	Yes No
Is there any large furniture?	Yes No	Have you called your insurance agent?	Yes No
Is there anyone with allergies?	Yes No	Have you called a repair person?	Yes No
Have called the health inspector?	Yes No	Are you still living in structure?	Yes No

Notes:

Cause of loss:

Areas Affected: Which areas and the type of flooring affected: **C**-Carpet, **V**-Vinyl, **T**-Tile, **H**-Hardwood, **O**-Other

Hall:	Liv Room:	Rec Room:	Dining:	Kitchen:	Bath:
Hall:	Study:	Den:	Master Bed:	Bedroom:	Bedroom:

Other areas affected:

General and Payment Information

Age of the structure: Style:

Steps already taken:

Special concerns of customer:

Has anyone already inspected the situation: Yes No Name: Title:

How will you handle the emergency fee:

Marketing Information

Where did you get our phone number?

How did you first learn about our company?

Referring Company: Referring Individual:

Instructions to Customer:

Emergency Service Response:

Technician's Notified Name:	Time:	By:
Technician's Arriving Name:	Time:	By:

Customer Briefing / Work Authorization

Instructions and authorization for the restoration of your property

Initial Please
⇓

☐ **Effect of Abnormal Water:** Water indoors is an abnormal condition and can cause or contribute to a number of problems. The damage and health implications increase the longer materials remain wet. There are at least four progressive stages of indoor water damage as follows:

1. Water migrates into areas not originally affected. Water spreads laterally into adjoining rooms, penetrates materials below and wicks up into porous materials above. It is best to evaluate all affected areas, not just areas with visible saturation.
2. Saturated materials begin swelling as they absorb moisture. In many situations drying and restoration will restore them, and in others the damage may be permanent. Rapid drying will help reduce the chance of permanent damage.
3. As water evaporates it causes an abnormal humidity condition. High humidity can damage some porous materials if allowed to continue. Humidity damage is most common when the indoor humidity exceeds 60% over time. It is best to reduce the indoor humidity quickly to help prevent humidity damage.
4. After water intrusion, mold and bacterial spores germinate and multiply. Given acceptable growth conditions, one organism can multiply to more than one billion organisms in less than 24 hours. Microorganisms can cause mold damage and degrade indoor air quality. Some microorganisms produce antigens and allergens which cause allergies while others can cause infections and toxic diseases.

☐ **Reducing mildew growth & damage:** I understand that several things that can be done to reduce mold growth and damage to affected materials. (1) Water can be removed as quickly and as thoroughly as possible. (2) Antimicrobial agents can be applied to inhibit the growth of mold and other bio-contaminants. (3) Rapid air-movement can be introduced into areas and cavities to increase the rate of evaporation. (4) The relative humidity indoors can be reduced and then maintained between 25% and 45% RH if possible.

☐ **Authorization:** I the Owner/Agent for the job site listed below, authorize The Company to enter my property, furnish materials, supply all equipment and perform all labor necessary to preserve and protect my property from further damage, and to perform all restoration procedures necessary to repair and restore the carpet, furniture, structure and other furnishings.

☐ **Prices:** I understand that water damage is a progressive condition and that drying time varies depending on the types of materials, quantity of water, degree of saturation, airflow volume and velocity, temperature and the indoor and outdoor humidity. Therefore, I understand it is impractical to give an accurate quote for services before completion. I have been supplied with The Company's standard price list and agree to pay the prices listed. The Company agrees to keep accurate records and provide documentation if requested.

☐ **Stop Work-Hold Harmless:** In the event that The Company is not allowed to perform its recommended procedures and/or drying equipment is removed prematurely, I agree to release and hold The Company harmless, and indemnify The Company against any claims or actions that may result from such incomplete procedures.

(continued on next page)

☐ **Authorize Antimicrobials Agents:** I understand that in the best judgment of The Company all wet materials should be treated with a commercial antimicrobial agent to inhibit the growth of microorganisms during the drying process. I have received oral and written information regarding the antimicrobial agent and agree that all wet materials should be treated with this product as part of the restoration process. I understand that it is beyond the expertise of The Company to determine if someone may be sensitive to its application and will hold The Company harmless for its use.

☐ **No Antimicrobial Agents:** I have decided that **I do not want antimicrobial agents applied** to affected materials in my home/business. I will hold The Company harmless for any biological odors or growth, indoor air quality degradation and any other health concerns or damage that may occur due to microbiological activity during and/or after the restoration process.

☐ **Drying Equipment:** I understand that high velocity airmovers (fans) and dehumidifiers will be installed to increase the rate of drying. I will not allow children to play with this equipment and will not turn off the equipment without first calling The Company. I have been informed that the target humidity for my home/business is _____% and it may take several days to attain this condition. I will not leave windows open unless instructed by The Company, as this may retard the drying process. I will minimize entering affected rooms, especially those where carpet is being dried.

☐ **Safety:** If dehumidifiers or airmovers must be moved, they must be shut off and unplugged as it may be hazardous to move these units while they are operating. Exposed tackless strip is a danger even when covered; I will take care when walking near tackless strip. The floors may be slippery when wet; I will take extreme care if walking on or from wet flooring materials.

☐ **Equipment Responsibility:** I understand that I am responsible for loss or theft of drying equipment while in my care and custody and will take reasonable precautions to insure that this drying equipment is not lost or stolen.

☐ **Insurance Payments:** If payment is made directly to the Owner/Agent by the insurance company, it shall be endorsed over to The Company within three working days. Any and all charges for services not reimbursed by an insurance company are the sole responsibility of the Owner/Agent and are to be paid upon completion of the work. Payment terms to The Company are Net-30 days and late charges of 1.5% monthly are charged on any unpaid balance. We request direct payment from Insurance Company.

☐ **Other Costs:** As Owner/Agent of the property, it is understood that I have authorized the work and accept responsibility to The Company for services rendered. It is fully understood and agreed that the Owner/Agent is personally responsible for any and all costs, charges, deductible or depreciation not covered by insurance.

I have read and understand the information above and have received a copy for my records.

Location: ____

Owner/Agent: ______________________________________ **Date:** _______________

The Company: ______________________________________ **Title:** _______________

Draft copy of Customer Briefing / Work Authorization form for discussion and debate.
Forms of this type should be evaluated with an attorney.

FIRST PARTY WORK AUTHORIZATION

This authorization is made this _______ day of ______________, 20____ by and between YOUR COMPANY NAME (Y.C.N.), hereinafter referred to as (Y.C.N.) and HOMEOWNER/INSURED (NAME) presently residing at HOMEOWNER/INSURED'S BILLING ADDRESS hereinafter referred to as the **CUSTOMER.**

The **CUSTOMER** authorizes (Y.C.N.) to proceed with its recommended procedures to preserve, protect and secure from further damage the property located at the ADDRESS OF LOSS (SAME AS ABOVE IF APPLICABLE) and, providing the **CUSTOMER** has valid effective insurance coverage for all or part of the services to be performed by (Y.C.N.), **the CUSTOMER further authorizes and directs their insurance carrier to pay** (Y.C.N.) **direct, and to name** (Y.C.N.) **on any and all insurance drafts applicable to this loss.** (Y.C.N.) shall bill all charges and/or costs directly to the **CUSTOMER** (insured), and, as a courtesy only, a copy of these invoices will be mailed to the insurance carrier. It is fully understood and agreed to by the **CUSTOMER** that any and all charges and costs are due upon completion of work. It is fully understood that the **CUSTOMER** is personally responsible for any and all deductible, depreciation or any charges or costs not covered by insurance. Any and all charges for services not reimbursed by an insurance carrier are the sole responsibility of the **CUSTOMER** and are to be paid upon completion of work. Any exceptions must be approved by (Y.C.N.) General Manager, and a finance charge of 1.5% per month (minimum of $1.00), will be applied to any unpaid balance after thirty (30) days. The liability of (Y.C.N.) shall not be responsible for mysterious disappearance of any personal property or contents. In the event any legal proceedings must be instituted to recover the amount due, (Y.C.N.) shall be entitled to recover the cost of collection including reasonable attorney's fees.

EXECUTED AT NAME OF COUNTY, STATE, on the day and year first above written

Authorized Signature: ______________________________
(Insured or Acting Agent)

Print Name: ______________________________
Title: HOMEOWNER/INSURED
Date: ______________________________
Policy/Claim #: CIRCLE APPROPRIATE ONE
Our File #: YOUR JOB #

YOUR COMPANY NAME

Signature of Representative of your firm ______________
Title: ______________________________

NOTES ABOUT FIRST PARTY WORK AUTHORIZATION

1) Your firm should always retain original copy.

2) Always give the customer a copy after executing the work authorization.

3) Send copy to the agent, adjuster, or paying party (if not the insured).

4) You may want to highlight the **"DIRECTION OF PAYMENT CLAUSE"** with a highlighting felt marker on the copy you send to the paying party. If you type the statement below on the front of your invoice, this will often help in accelerating payment and/or ensuring your name is put on the draft.

> **NOTE: We formally request direct payment and/or inclusion of our name on any applicable drafts as assigned by your insured.**

Recommended Procedures for Executing Work Authorizations for Absentee Owner:

If there is a concerned responsible party with proof of power of attorney (POA), then they may be able to act as the legal agent to execute the work authorization.

In out-of-town ownership situations, when no P.O.A./agent exists, you may verbally execute a work authorization in the following manner:

1) Identify your company and describe the situation to the owner.

2) Explain what is required to proceed with the necessary steps to preserve and/or restore their property.

3) Read your work authorization and ask their verbal consent to proceed.

4) Ask them to instruct someone to act as their agent to execute the authorization.

5) Inquire if they wish to have their copy of the work authorization mailed to them or left with the assigned agent.

The above procedure may be performed by the FAX process when available to both parties.

When the FAX option is not available, five (5) items need to be documented in a phone log "Diary Type" book. This will protect your firm, in the event there is a misunderstanding at some phase in the job.

WRITE DOWN:

1) Date of the phone call

2) Time of the phone call

3) Full name of the person you talked to

4) Brief description of the conversation (i.e. decisions made and/or agreed to orally)

5) Signature of person taking the information

NOTE: The original copy should be removed from the log book, and stapled to the back of the original copy of your work authorization, for filing purposes. The carbon (NCR copy) permanently remains in the log book. The filled-up books should be filed with your archives. In the event litigation is involved, the log book may be used to illustrate the sequence in which you documented, and entered the data from your conversation. Should it become necessary for legal proceedings, the log book may be submitted as evidence to verify the accuracy of your version of the sequence of events.

> **THE JOB SHOULD NOT COMMENCE UNTIL A RESPONSIBLE PARTY HAS SIGNED *YOUR* WORK AUTHORIZATION!**

Third Party Work Authorization

This authorization is made this _______ day of ____________, 20____ by and between **YOUR COMPANY NAME (Y.C.N.)** hereinafter referred to as **(Y.C.N.)**, and **PROPERTY OWNER/CLAIMANT** presently residing at **PROPERTY OWNER/CLAIMANT'S BILLING ADDRESS** herein after referred to as **CUSTOMER.**

The **CUSTOMER** authorizes **Y.C.N.** to proceed with its recommended procedures to preserve, protect and secure from further damage the property located at **LOSS ADDRESS OR SAME AS ABOVE, IF APPLICABLE** and with the understanding that **NAME OF LIABLE THIRD PARTY** is hereby responsible for all services and charges to be performed by **Y.C.N.**, as authorized by **NAME OF PERSON REPRESENTING THIRD PARTY**, and the **CUSTOMER** further authorizes and directs **NAME OF LIABLE/ THIRD PARTY** to pay **Y.C.N.** direct.

It is fully understood that the **CUSTOMER** is personally responsible for all charges, deductible, and depreciation not covered by **NAME OF LIABLE THIRD PARTY**. Any exceptions must be approved by **Y.C.N.** General Manager. The liability of **Y.C.N.** is expressly limited to the total amount of the services authorized herein and in no event shall **Y.C.N.** its agents or assigns, be liable for consequential damages of any kind. **Y.C.N.** shall not be responsible for any mysterious disappearances of any personal property or contents. In the event any legal proceedings must be instituted **Y.C.N.** shall be entitled to recover the cost of collection including reasonable attorney's fees. All charges and costs are due upon completion of work. Late charges of 1.5% per month (minimum of $1.00) will be charged on any unpaid balance after thirty (30) days.

Executed at **NAME OF COUNTY, STATE** on the day and year first above written.

Authorized Signature: ___________________________________
(Claimant or Acting Agent)
Print Name: ___________________________________
Title: **PROPERTY OWNER/CLAIMANT**_______________
Policy/Claim #: ___________________________________
Purchase Order #: **IF APPLICABLE** _________________
Our File #: **YOUR JOB NUMBER** __________________

YOUR COMPANY NAME
REPRESENTATIVE OF YOUR FIRM_________________
Title: ______________________________________
Signature of: **NAME OF PERSON FROM THIRD PARTY**
Representative: **THEIR SIGNATURE IF AVAILABLE** __

Notes About the Third Party Work Authorization

1. Be sure to retain the original.
2. Issue a copy to signing party (claimant) and third party representative (if applicable).
3. Send a copy to liable/third party with your invoice if they did not receive one at the job site.
4. Highlight “Direct Payment” clause below.
5. Note on your invoice:

WE REQUEST DIRECT PAYMENT AS ASSIGNED BY THE CLAIMANT

Your Company Name
Address
Phone Number

Work Authorization Form

Customer Name: ________________________________

Address: ____

City, Zip, State: ________________________

I authorize _______________________________________ to perform the following work:

____ ___

____

as approved for payment by: _______________________________for the purpose of treating:

____

as soon as possible.

As the owner of the this property, I understand that I must authorize this work and accept responsibility for payment to ______________________________________ (Company) upon completion, and agree that if their bill is not so paid to cover any collection costs incurred including attorney's fees up to 33½% plus 2% monthly service charge applicable 30 days after completion.

I authorize my insurance company: ____ to make direct payment to ____________________________________ (Company) for doing this work and request that their name be included on any check or draft issued to me consequent to my insurance claim. I also understand that I am responsible for my deductible in the amount of ____________.

Date: _____________________________

Signed: ___________________________

Signed: ___________________________

Your Company Name

Address
Phone Number

Work Authorization and Assignment of Insurance Benefits

ASSIGNMENT:

I hereby assign to ______________________________(Company) and authorize my insurance company to pay directly to ______________________________(Company) such portion of the proceeds of my insurance policy as shall be required to fully pay ______________________________(Company) for their restoration services rendered pursuant to this work authorization.

I certify that I authorize ______________________________(Company) to do the work ______________________________(Company) deems necessary to complete restoration, and I understand that I am responsible for payment of my deductible and for any and all charges which are not paid by my insurance company, and agree to pay or direct payment to ______________________________(Company) upon receipt of their invoice. I also agree that any fees for attorneys or any collection process required for non-payment of the balance of the invoice shall be paid to ______________________________(Company) in addition to the 1 ½% monthly service charge applicable thirty days after date of invoice on the unpaid balance of said invoice.

I also authorize ______________________________(Company) to supply information regarding this claim to my insurance company and to make a full report of all work done by ______________________________(Company).

DISCLAIMER:

Customer acknowledges that when ______________________________(Company) attempts to restore items which were damaged by fire, wind, water, vandalism, or other traumatic occurrence, that permanently discolored, faded and/or bleached areas might remain even though ______________________________(Company) covenants to use every diligent effort and means available to remove all spots. Customer understands that there is no guarantee that in all circumstances that items can be restored to their condition prior to the disaster. In the case of items made of, or covered with, fabric, there may be shrinkage, fugitive colors, fabric separation, changes in texture and other hidden conditions, such as loose carpet or tackless strip, improperly laid carpet, broken seams, delamination, or permanent deterioration that cannot reasonably be anticipated prior to restoration or cleaning attempts. In the event that these items cannot be restored to pre-disaster condition, normal recourse is to file an additional damage claim under the appropriate insurance policy provisions. ______________________________(Company) will not be held responsible for such conditions. Any corrections made because of the above named conditions are the owner's responsibility together with any cost involved in correcting these conditions. Initials _____

Customer further acknowledges receipt of Letter of Introduction Initials _____

AUTHORIZATION:

Authorization to begin work hereby granted: ______________________________ Date Signed ____/____/____
Signature

Insured ______________________________ Address where work is done ______________________________

______________________________ Amount of Deductible/Partial Payment ______________________________

Deductible/Partial Payment Paid ☐ Yes ☐ No ☐ Check ☐ Cash Check No. ______________________________ Date ____/____/____

Date of Loss ____/____/____

Policy No. ______________________________ Claim No. ______________________________ P.P. No. ______________________________

Contact Person ______________________________ Address ______________________________ Phone ______________________________

Work Inspected and Accepted by ______________________________ Date ____/____/____

Your Cleaning & Restoration Company

123 Any Street
Any City, USA 12345
(800) 555-2121

Client:	Insurance Company:
Address:	Agent:
City:	Adjuster:
State:Zip:	Policy #:
Home Phone #:	Claim #:
Business Phone #:	Deductible:
Date of Loss:	Type of Loss:

WORK ORDER AGREEMENT TO PERFORM EMERGENCY SERVICES AND/OR REPAIRS AND DIRECT PAY AUTHORIZATION

This agreement is made between The Client listed above (hereinafter referred to as The Client) and Your Company (hereinafter referred to as The Contractor). The Client has suffered a sudden and unexpected loss to property due to

The Client hereby authorizes The Contractor to perform the following services to midigate the loss and/or maintain a suitable living condition and/or comfort level:

The Client certifies that the damaged property has the appropriate insurance coverage to cover this loss and that The Client is responsible for payment of any deductible as well as any charges on the final billing for this service not covered by The Client's insurance policy or not paid by The Client's insurance company for any reason.

The Client further authorizes and instructs The Client's insurance company to pay directly to The Contractor (or to include The Contractor as a co-payee on the check or draft) the amount shown on the final billing for work done by The Contractor in connection with this claim. The Client also understands that the insurance company is billed as a courtesy and convenience to The Client. Should The Client's Insurance Company fail to honor this agreement, The Client will pay the Contractor any balance due from The Client's personal funds on the tenth (10th) calendar day from the date of this agreement

It is The Client's complete understanding that The Contractor is working for The Client and not The Client's insurance company. Therefore, it is understood that The Client is ultimately responsible to The Contractor for payment of said services.

The Client also authorizes The Contractor to supply information regarding this claim to The Client's insurance company and to make a full report of all work done by The Contractor.

The Client has read the above and understands that this form, constitutes a contract for services performed. Should it become necessary to institute legal proceedings to enforce collection of this contract, the prevailing party shall be entitled to reasonable attorney fees.

WORK ORDERED BY: ____

Date: ________________________________

The person ordering the above work will be held responsible for payment.

CONTRACTOR ____

REPRESENTATIVE: ____

Date: ________________________________

Initial Inspection and Loss Assessment

Initial Emergency Steps to Preserve and Protect the Property from Further Damage

Communication and Safety

_____ Arrive at the job within one hour of the initial call

_____ Ensure that you have initiated the Call Report form

_____ Ensure that you have a signed work authorization form

_____ Collect the emergency fee or the insurance deductible

_____ Determine and help contain the source of the water damage

_____ Consider hazards and help ensure the safety of the occupants

_____ Communicate the procedures and timetable for restoration to all parties

_____ Complete the Call Report Form and initiate documentation of affected materials

Initial Inspection and Assessment

_____ Evaluate horizontal moisture migration from the source (areas affected)

_____ Evaluate the vertical moisture migration and the height of wicking up walls

_____ Evaluate the saturation or MC into and under structural materials like flooring

_____ Evaluate primary moisture in cabinets, bookcases and built-in furniture

_____ Evaluate primary moisture content and/or damage to contents (furniture)

_____ Evaluate existing and potential secondary damage to contents and structure

_____ Remove, relocate or protect sensitive materials from secondary damage

_____ Determine the types of materials that have been exposed to moisture

_____ Determine which materials will most likely incur permanent water damage

_____ Determine if the carpet and cushion will be replaced or dried on-site

_____ Evaluate the air conveyance system (HVAC) for water and/or contamination

_____ Note pre-existing conditions (urine, mold) and point these out to customer

_____ Check and record the humidity/temperature outdoors and indoors

Initial Procedures

_____ Pump out standing water as required in basements and crawl spaces (cavities)

_____ Extract the moisture from highly absorbent materials like carpets, and upholstery

_____ Initiate cushion compression procedures as required for on-site cushion drying

_____ Remove structural materials (drywall, cushion, etc.) that will not be restorable

_____ Remove cove base or baseboard as required to increase the rate of wall drying

_____ Ventilate or dehumidify wall and floor cavities to increase rate of evaporation

_____ Remove wet insulation that has been damaged by water (R-factor/organic issues)

_____ Treat affected materials with antimicrobials as required to reduce odors/damage

_____ Remove contents, or plan on-site work for contents that will require special handling

Drying Procedures

_____ Evaluate quantity of drying equipment needed such as airmovers and dehumidifiers

_____ Install dehumidifiers as required to maintain/reduce humidity below 45% RH

_____ Install TurboDryers® as required to increase the rate of evaporation

_____ Install specialty blowers as required to dry walls and hardwood flooring

_____ Communicate your drying plan and timetable with all involved parties

_____ Record humidity and temperature indoors and outdoors before leaving

Health Information and Consent to Proceed Form

In order to provide you with the best possible service, we request that you provide us with the following information:

1. I agree_______ decline_______ to provide the following information. (Please initial.)

2. Does any resident have any known allergies? Yes_____ No_____
   Please list:

   ____

   ____

3. Does any resident have any sensitivity to chemicals? Yes_____ No_____
   If so, please list:

   ____

   ____

4. Is any resident under the care of a physician? Yes _____ No_____

   Name of resident(s) under care: ________________________________

   Physician's Name: __________________ Telephone: ________________

5. Are there any residents under the age of 6 years old? Yes ____ No ____

6. Are there any residents over the age of 60 years old? Yes ____ No ____

7. Are there any residents with respiratory problems? Yes ____ No ____

8. Are there any residents that have a deficient immune system? Yes ____ No ____

9. Please list any concerns you may have about our services or about your health:

____

____

____

I have read the information provided to me and have reviewed or been offered information on the MSDS (Material Safety Data Sheets) regarding the chemicals that may be used in my residence and I hereby give my consent for necessary services to be performed.

____

Owner / Occupant Date City

Acknowledgment of Risks and Release

I) Recitals

Whereas, ___________________________ ("customer") desires ______________________________
(Company Name)
(hereinafter collectively referred to as the "Services") to property; and

Whereas, in order to effectively perform the Services, it is necessary for ________________________, to use cleaning agents, deodorizers, and chemicals (collectively referred to as the "Chemicals"); and

Whereas, some individuals are sensitive to the chemicals and may become sick or nauseated or suffer potential risks from being exposed to such Chemicals; and

Now, Therefore, in consideration of the mutual covenants and conditions contained herein, the parties agree as follows:

II) Terms and Conditions

1. **Notification of Risks.** Customer hereby acknowledges that _________________________(Company) has advised Customer that the Customer may suffer potential health risks as a result of the application of the Chemicals used in performing the Services. Customer hereby acknowledges that _________________________(Company) has provided him/her with a copy of the Material Safety Data Sheets for the Chemicals to be used in performing the Services. Customer further acknowledges and understands that the Material Safety Data Sheets are not prepared by _____________________________ (Company) but, rather, are prepared by the manufacturers of the Chemicals and that the health risks listed on such safety sheets may not be all inclusive.
2. **Opportunity to Seek Medical Opinion.** Customer hereby represents that he/she has had an opportunity to seek a medical opinion on the potential health effects concerning the Chemicals to be applied during the Services. Customer further represents that he/she accepts the risks as set forth herein.
3. **Release.** Customer hereby releases and forever discharges __________________________(Company) and it's agents, principals, owners, shareholders, directors, managers, and employees from any and all claims as to injuries, damages, and health problems of any nature, whether known or unknown, suffered as a result of the application of the Chemicals.

The undersigned hereby declares that the terms of the Acknowledgment of Risks and Release have been completely read and are fully understood and voluntarily accepted.

Dated this _______ day of ___________________, 20______.

____
Customer Signature

Record of Drying Conditions

Job Name ______________________ Area: ______________________

Address ______________________ Time Area Affected: ____________

Phone ______________________ Class: 1 2 3 X

Inspection Date/Time	Temp	Rh	G.P.P.	Temp	Rh	G.P.P.	Temp	Rh	G.P.P.	Temp	Rh	G.P.P.	Temp	Rh	G.P.P.	Temp	Rh	G.P.P.	Temp	Rh	G.P.P.

Inspection Date/Time	Temp	Rh	G.P.P.	Temp	Rh	G.P.P.	Temp	Rh	G.P.P.	Temp	Rh	G.P.P.	Temp	Rh	G.P.P.	Temp	Rh	G.P.P.	Temp	Rh	G.P.P.

Record of Moisture Levels

Customer:	Date of Loss:
Street:	Today's Date:
City:	Technician:

Meter:				
Scale:				
Location:				

	A	B	C	D	E	F	G	H	I	J	K	L	AVG
Day 1													
Day 2													
Day 3													
Day 4													
Day 5													

DETERMINING MOISTURE CONTENT OF AIR IN A BUILDING

1. Determine dry-bulb temperature and relative humidity of the building by use of your thermo-hygrometer.

   Temperature _________ Relative Humidity ________ %

2. Refer to your psychrometric chart or calculator. Locate dry-bulb temperature on the bottom line. Draw a line straight up to the RH% curve line on the chart. At the intersection, draw a straight line over to the right side. The numbers reading up are the grains per pound of dry air (GPP).

   Grains per pound is ____________

3. To figure weight of air in the building, divide the cubic feet of the building by fourteen. (14 cubic feet = 1 pound of dry air)

   Cubic feet _______ cu. ft. ÷ 14 = _______ Total lbs. of air

4. Multiply the grains per pound (GPP) from figures above (2.) times the total pounds of air (3.). This is the total grains of moisture in all the air at that time.

   GPP ______ x total lbs. ______ = ______ Total grains

5. Divide total grains in the air by 7000. (7000 grains = 1 pound of water). This gives the total amount of moisture that is present in pounds. (Pounds of water)

   Total grains in air ________ ÷ 7000 = ______ Total Pounds

6. Divide the total pounds present by 8.34 to get total gallons of water present. (1 gallon of water = 8.34 lbs)

   Total Pounds _________÷ 8.34 = __________ Total Gallons

7. Multiply total gallons by 8 to get the total pints of water

   Total Gallons _________X 8 = __________ Total Pints

EMC Standards Chart

Dri-Eaz Flood House

MATERIALS DRY TESTED ***In a "Dry" Normal Environment***	**Meter**	**Meter**	**Meter**	**Meter**	**Meter**
Baseboard, wood					
Oriented strand board (OSB)					
Concrete slab floor (w/o sealer)					
Concrete blocks					
Drywall, painted					
Insulation, fiberglass batts					
Plaster wall (skim coated on drywall)					
Plywood					
Oak hardwood flooring					
2X6 Bottom plate (treated lumber)					
Underlay (under resiliant flooring)					
Toe kicks (treated lumber)					
Kiln dried dimensional lumber					
Engineered lumber					

Note: **These readings represent several samples of each material. Reading variations are normal.**

Wood materials are considered dry enough when readings are within 4% of EMC (equilibrium moisture content), according to IICRC S500

Measurements may vary from meter to meter.

Wood Equilibrium Moisture Content

The moisture content of wood below the fiber saturation point is a function of both the relative humidity and temperature of the surrounding air (vapor pressure).

Temperature Dry Bulb F							
RH	**30° F** (-1° C)	**40° F** (4° C)	**50° F** (10° C)	**60° F** (16° C)	**70° F** (21° C)	**80° F** (27° C)	**90°F** (32° C)
20%	4.6	4.6	4.6	4.6	4.5	4.4	4.3
25%	5.5	5.5	5.5	5.4	5.4	5.3	5.1
30%	6.3	6.3	6.3	6.2	6.2	6.1	5.9
35%	7.1	7.1	7.1	7.0	6.9	6.8	6.7
40%	7.9	7.9	7.9	7.8	**7.7**	7.6	7.4
45%	8.7	8.7	8.7	8.6	8.5	8.3	8.1
50%	9.5	9.5	9.5	9.4	9.2	9.1	8.9
55%	10.4	10.4	10.3	10.2	10.1	9.9	9.7
60%	11.3	11.3	11.2	11.1	11.0	10.8	10.5
65%	12.4	12.3	12.3	12.1	12.0	11.7	11.5
70%	13.5	13.5	13.4	13.3	13.1	12.9	12.6
75%	14.9	14.9	14.8	14.6	14.4	14.2	13.9
80%	16.5	16.5	16.4	16.2	16.0	15.7	15.4
85%	18.5	18.5	18.4	18.2	17.9	17.7	17.3
90%	21.0	21.0	20.9	20.7	20.5	20.2	19.8
95%	24.3	24.3	24.3	24.1	23.9	23.6	23.3

The equilibrium moisture content is defined as that moisture content at which the wood is neither gaining nor losing moisture. When this is achieved, an equilibrium condition has been reached.

Water Removal Estimator

Estimated Pint Removal Based on Grain Depression and CFM

	DrizAir Refrigerant DHs					Desiccants			HVAC Systems				
Model	**Evo**	**1200**	**120**	**2000**	**2400**	**DT150**	**Pro150**	**DT325**	**1-ton**	**2-ton**	**3-ton**	**4-ton**	**5-ton**
CFM	**160**	**227**	**250**	**360**	**360**	**110**	**88**	**250**	**600**	**1200**	**1800**	**2400**	**3000**
Grain Depression (in GPP)													
1	2	3	4	5	5	2	1	4	8	17	25	34	42
2	5	6	7	9	10	3	2	7	17	34	51	68	85
3	7	10	11	14	15	5	4	11	25	51	76	101	127
4	9	13	14	18	20	6	5	14	34	68	101	135	169
5	11	16	18	23	26	8	6	18	42	85	127	169	211
6	14	19	21	27	31	9	7	21	51	101	152	203	254
7	16	22	25	32	36	11	9	25	59	118	178	237	296
8	18	26	28	36	41	12	10	28	68	135	203	271	338
9	20	29	32	41	46	14	11	32	76	152	228	304	381
10	23	32	35	45	51	16	12	35	85	169	254	338	423
11	25	35	39	50	56	17	14	39	93	186	279	372	465
12	27	38	42	54	61	19	15	42	101	203	304	406	507
13	29	42	46	59	66	20	16	46	110	220	330	440	550
14	32	45	49	63	71	22	17	49	118	237	355	474	592
15	24	48	53	68	77	23	19	53	127	254	381	507	634
16	26	51	56	72	82	25	20	56	135	271	406	541	677
17	28	54	60	77	87	26	21	60	144	288	431	575	719
18	41	58	63	81	92	28	22	63	152	304	457	609	761
19	43	61	67	86	97	29	24	67	161	321	482	643	803
20	45	64	70	90	102	31	25	70	169	338	507	677	846
21	47	67	74	95	107	33	26	74	178	355	533	710	888
22	50	70	78	99	112	34	27	78	186	372	558	744	930
23	52	74	81	104	117	36	29	81	195	389	584	778	973
24	54	77	85	108	122	37	30	85	203	406	609	812	1015
25	56	80	88	113	128	39	31	88	211	423	634	846	1057
26	59	83	92	117	133	40	32	92	220	440	660	880	1099
27	61	86	95	122	138	42	33	95	228	457	685	913	1142
28	63	90	99	126	143	43	35	99	237	474	710	947	1184
29	65	93	102	131	148	45	36	102	245	491	736	981	1226
30	68	96	106	135	153	47	37	106	254	507	761	1015	1269

Formula for Estimating the Rate of Water Removal

Begin with CFM of Equipment			CFM (Cubic Feet processed per minute)
Multiply by 60 (CFM x 60)		=	CFH (Cubic Feet processed per hour)
Multiply by 24 (CFH x 24)		=	CFD (Cubic Feet per Day)
Divide CFD by 14 (CFD / 14)		=	Pounds of Dry Air processed per day
Grain Depression (GD)		=	GD (Ambient GPP minus Equip. Output GPP)
Multiply GD by Pounds of Dry Air		=	Total Grains of Moisture Removed per Day
Divide Total Grains Removed by 7000		=	Pounds of Water Removed per Day
Divide Pounds of Water Removed by 8.34		=	Gallons of Water Removed per Day
Multiply Gallons of Water Removed by 8		=	Estimated Rate of Removal in Pints/Day

This Calculation Chart Estimates the Rate of Water Removal in Pints per Day at a Specific Point in Time

Centigrade

PSYCHROMETRIC CHART

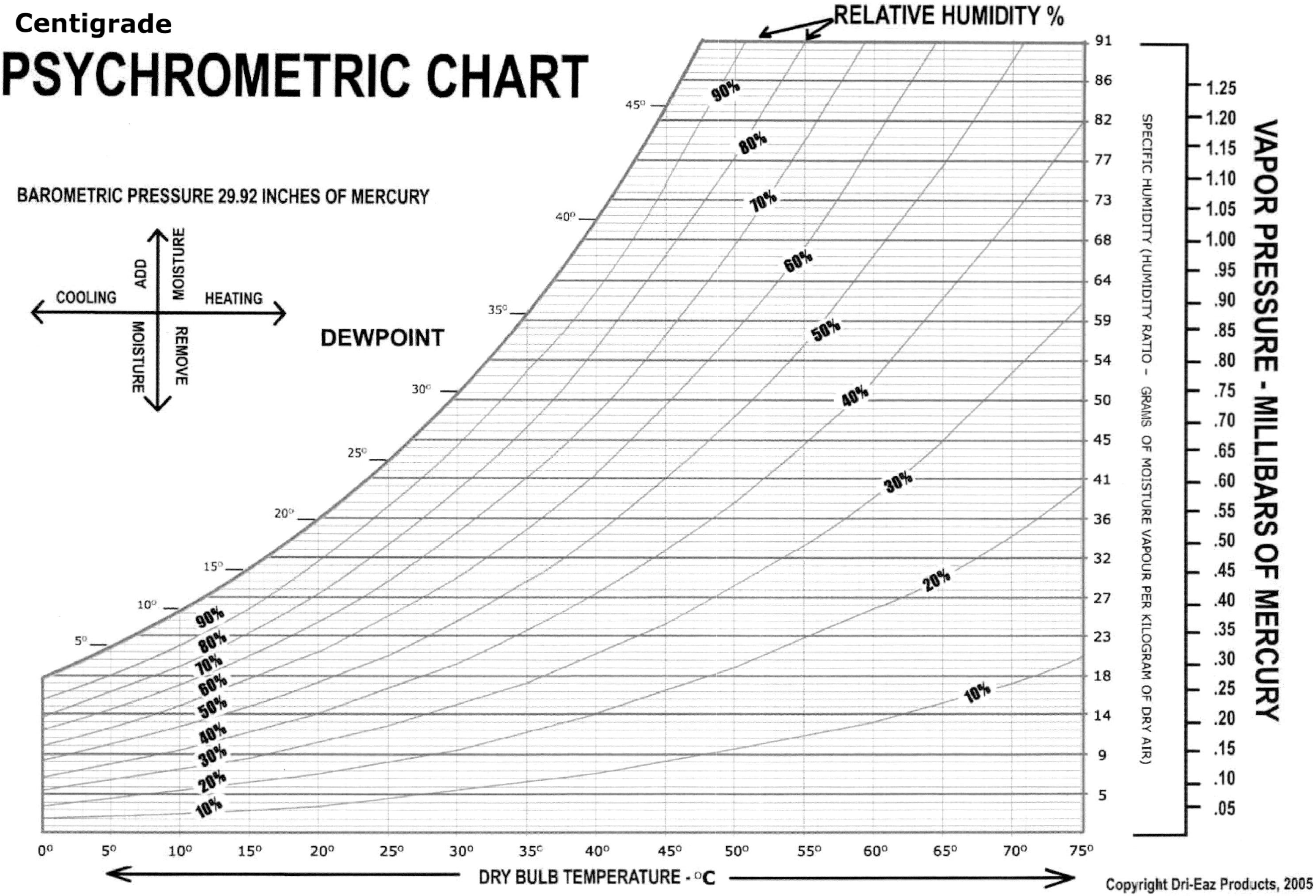

Fahrenheit

PSYCHROMETRIC CHART

BAROMETRIC PRESSURE 29.92 INCHES OF MERCURY

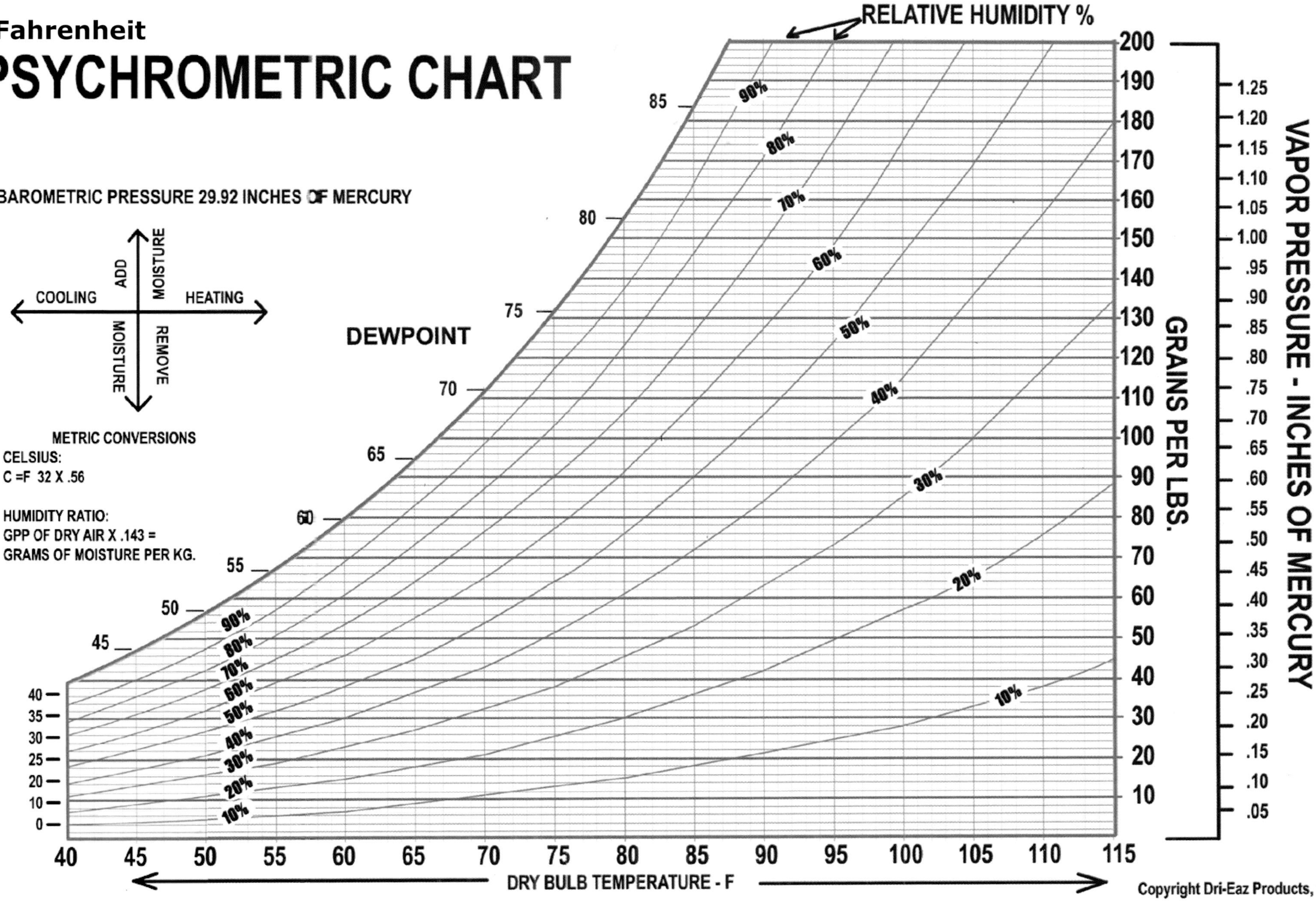

METRIC CONVERSIONS

CELSIUS:
C =F 32 X .56

HUMIDITY RATIO:
GPP OF DRY AIR X .143 =
GRAMS OF MOISTURE PER KG.

Acknowledgment of Satisfactory Completion

Insured/Claimant: ____

Loss Address: ____

Policy #: _______________________________ Purchase Order #: ______________________

Claim #: _______________________________ Our File #: _____________________________

This is to certify that the Restoration Services provided by _________(Y.C.N.)________ at the above mentioned property have been completed to my/our satisfaction. These services were necessitated by a ___________________________________ damage loss suffered on ________________.

(Type of Loss) Month Day Year

Upon execution of this certificate and payment of all related invoices, ______(Y.C.N.)_________will guarantee all restoration workmanship for a period of 12 months from the below stated date.

Authorized Signature: _______________________________________
(Insured/Claimant or Acting Agent)

Print Name: _______________________________________

Title: _______________________________________

Date: _______________________________________

(YOUR COMPANY NAME)

____
Signature of Representative of Your Firm

Title: _______________________________________

Glossary

A

ACGIH: American Council of Governmental and Industrial Hygienists. Makes recommendations and sets standards for limits of contaminants in water and air.

Acknowledgement of Satisfactory Completion: A document containing a statement of satisfactory completion that is signed by the property owner or other authorized entity; completed after the project has been finished. Also see the Appendix for a sample of this form.

AEH: Air Exchanges per Hour. Refers to the rate at which air is exchanged between two spaces.

AFD: See *Air Filtration Device.*

Affected Area: Room or rooms that contain materials adversely affected by abnormal water intrusion into a space.

Aggressive Drying: Term used to describe restorative drying approaches that minimize material removal through the use of drying equipment and frequent monitoring. Also referred to as *In Place Drying.*

AHAM: Association of Home Appliance Manufacturers. Publishes test results for water removal of dehumidifiers at conditions of 80°F/60%RH (27°C/60%RH) over 24 hours.

Air Exchange Heat Drying System: A drying system that exchanges indoor, moisture laden air with heated outdoor air. See also *Open Drying System.*

Air Exchanger: Device that exchanges outside air for inside air while saving heat (or cool air, in summer). Can be used to create a combination drying system. Does not cause pressure differentials because airflow *in* equals airflow *out.*

Air Exchanges Per Hour: See *AEH.*

Air Filtration Device: Machine with several filters, including a HEPA filter, designed to remove contaminants from the air. Can be set up to negatively or positively pressurize an area or to clean air in a space without pressurization.

Airflow: One of the three components of psychrometry that can be controlled, along with humidity and temperature. The role of airflow is to circulate dry air across wet surfaces, removing the moist boundary layer and accelerating evaporation.

Air Mover: Specialized blower – either carpet drying or axial fan – designed for creating airflow to dry structures.

Air Scrubber: See *Air Filtration Device.*

Alcohol: Colorless volatile flammable liquid, C2H5OH, used as a solvent and in drugs, cleaning solutions, explosives, and intoxicating beverages. Also called ethanol, ethyl alcohol, grain alcohol.

Aldehyde: A class of highly reactive organic compounds made by oxidizing alcohol. See *glutaraldehyde.*

Algae: A group of mostly aquatic, photosynthetic organisms that range from single celled microorganisms to large kelp.

American National Standards Institute: The American National Standards Institute (ANSI) is a private, non-profit organization (501(c)3) that administers and coordinates the U.S. voluntary standardization and conformity assessment system. ANSI is responsible for the development of ANSI Z117.1-1989, *Safety Requirements for Confined Spaces.*

Amp: See *Amperage.*

Amperage: Strength of an electric current measured in amperes (amps). Electrical circuits are designed and rated to carry a certain number of amps, and should not be loaded with more than 80% of that amperage. For example: a 15 amp circuit should not carry more than a 12 amp load.

ANSI: See *American National Standards Institute.*

Antimicrobial: Any agent that is capable of destroying or inhibiting the growth of microorganisms (i.e., a temperature, chemical or radiation).

Asbestos: An incombustible, flexible, fibrous mineral material made of impure magnesium silicate. Used for fire proofing. Inhaling fibers has been known to cause lung cancer (asbestosis). Restorers are required by United States Federal Law (29 CFR 1910.1001 and 29 CFR 1926.1101) to perform special procedures, including testing, before handling building materials that might contain asbestos.

ASHRAE: American Society of Heating, Refrigeration and Air Conditioning Engineers. Makes recommendations and sets standards for heating and cooling systems.

Aw: See *Water Activity.*

Awl: Carpet installation tool used to disengage carpet, in conjunction with a knee kicker. The awl tip is pushed through the primary and secondary backing of the carpet to disengage it from tackless strip.

Axial Fan: Air mover designed for the purpose of ventilation. Axial fans typically produce a high volume of air at relatively high pressure – critical for ventilation, especially when ducting is employed to carry the airflow.

B

Backing: The component of a carpet to which the face fiber material is attached, normally only visible from the bottom of the carpet. Many residential carpets are made with two backings, a primary backing to hold the face fiber and a secondary backing, added to give the carpet structural stability.

Bacteria: The plural for bacterium. *Bacteria* are a unicellular microorganism and may be saprophytic, parasitic and pathogenic. See *also endotoxins, exotoxins.*

Balanced Drying System: Refers to the proper sizing of dehumidification equipment or other humidity control. A *balanced drying system* ensures that water vapor is being removed (dehumidification) faster than it is being added (evaporation) to the air. The net result will be a gradual decrease in humidity over time.

Base Plate: Bottom of a wall in wood or metal frame construction. (Studs are attached to the base plate.) Also called *bottom plate, seal plate, or sole plate.*

Berber: Type of carpet characterized by loop pile. Loops can be level, multi-level, or a combination of multi-level and cut pile. Berber carpets most often contain olefin fibers, but can also be made of other fibers.

Biocide: Chemical poison designed to kill microorganisms or other organisms. Always use biocides in accordance with the directions on the label of the product, industry standards and local laws.

Black Water See *Category 3.*

Blocking: Setting furniture and other items on blocks of foam, plastic or wood to help prevent secondary damage and provide a space for airflow underneath the item.

Blood Borne Pathogen Plan: A written plan outlining required PPE, procedures and practices when blood is likely to be present in the work area. This plan is required by OSHA, mandated by the OSHA Blood Borne Pathogens Standard (1991).

Blower: An air moving device. Generally, the term refers only to air movers that employ blower wheels, not fan blades, to generate airflow, resulting in higher pressures.

Boroscope: A device that allows for visual inspection of otherwise inaccessible cavities through the use of flexible fiber optics.

Boundary Layer: Layer of moisture laden air that is in contact with a wet surface. If the surface is releasing moisture, the boundary layer can quickly become saturated, inhibiting further evaporation. Air movement's purpose is to sweep away the boundary layer.

Bound Water: Water contained in wood that has absorbed into the walls of the wood cells. Difficult to remove as compared with free water.

Brick: A rectangular block of clay dried in a kiln until hard, then used as a building material or paver.

BTU: British Thermal Unit – a specific measure of heat energy. Calculations using power consumption can determine approximate heat added to air: BTU per hour = watts x 3.4 or BTU per hour = amps x volts x 3.4.

C

C channel: Bottom plate of a metal stud wall that can hold quantities of water and which must be inspected and addressed.

CADR: Clean Air Delivery Rate (of an air scrubber). Measured by multiplying the air scrubber's CFM by its filter efficiency. See also HEPA.

Calibration: Checking or adjusting (against a known standard) the accuracy of an instrument used for measurement.

Call Report Form: A document designed to capture critical information during an initial phone conversation between the restorer and the property owner. Also see the Appendix for a sample of this form.

Capillary Action: The movement of water along the surface of a solid object that causes it to migrate through a material. Observed as an elevation of water at the edge of a container, such as a glass, compared to the center.

Capillary Tube: A component in a refrigeration system that restricts the flow of refrigerant, allowing for the separation of the condenser and evaporator coils and therefore generation of high and low temperatures. Also called a restriction device.

Carbon Monoxide: A colorless, highly poisonous gas (CO) created by the incomplete combustion of carbon or carbonaceous material (i.e., fossil fuels such as gasoline, diesel, propane).

Carpet: A general designation for a fabric used as a floor covering that is typically made from synthetic or natural fibers that are either woven through or adhered to a backing material. The term carpet is commonly used to describe floor coverings that cover an entire area of flooring, as opposed to movable, temporary coverings such as an area rug. See also *Woven Carpet, Stretched-In Carpet, Direct Glue Carpet.*

Carpet Awl: A device used during the removal of stretched-in carpet. It is shaped like an ice pick, with a small handle attached to a steel shaft that comes to a point. The tip of the *carpet awl* is forced through the carpet backing, so that the carpet can be lifted from the back.

Carpet Delamination: See *Delamination.*

Carpet Dryer: Drying tool that sends a high velocity stream of air across floor and wall surfaces. Has medium air pressure and a high volume of airflow.

Category 1 (Clean Water): Originates from a source that does not pose substantial threat to humans. May deteriorate in cleanliness when in contact with structural surfaces and content materials.

Category 2 (Gray Water): Water containing a significant degree of chemical, biological and/or physical contamination. Cushion (underlay) must be removed and replaced. Carpet that is saved must be cleaned with hot water extraction.

Category 3 (Black Water): Grossly unsanitary water, containing pathogenic agents arising from sewage or other contaminated water sources. Water originating from a source that poses substantial threat to human health. Carpet and cushion must be removed and disposed.

CE: Conformité Européenne. The European equivalent to Underwriter's Laboratory (UL). The CE mark on a product label indicates that the product complies with set European Health and Safety standards.

Cellulose: A complex carbohydrate, $(C_6H_{10}O_5)_n$, that forms the main element of the cell wall in most plants, and is important in the manufacture of numerous products, such as paper, textiles, pharmaceuticals, and explosives.

CFD: Cubic Feet per Day. A measure of airflow by volume. See also *CFM.*

CFH: Cubic Feet per Hour. A measure of airflow by volume. See also *CFM.*

CFM: Cubic Feet per Minute – the imperial measurement used for volume of airflow. Most often a measure of effectiveness in an air mover or axial fan. Can be used to determine the amount of ventilation in a structure. See also *CMH.*

CFR: See *Code of Federal Regulations.*

Checking: Secondary damage that can occur when a saturated material is dried too quickly, and the outer portion shrinks and splits around a swollen and saturated interior.

Chemical Material Safety Data Sheet: See *MSDS.*

Class of Water Loss: Refers to the rate of potential evaporation in a wet area. The *class of water loss* is determined by evaluating the amount of water in the affected area and the type of materials present.

Class 1 Water Loss: Water loss with a slow rate of evaporation, affecting only part of a room or area, with low-permeance/porosity materials (e.g., plywood, particle board, structural wood, VCT, concrete). Little or no wet carpet and/or cushion is present. Minimum moisture is absorbed by materials, which release that moisture slowly.

Class 2 Water Loss: Water loss with a fast rate of evaporation, affecting the entire room of carpet and cushion, where water has wicked up walls less than 24" (61cm). Moisture remains in structural materials (e.g., plywood, particle board, structural wood, concrete).

Class 3 Water Loss: Water loss with the fastest rate of evaporation, where carpet and pad have been affected by water from above or where water has wicked up walls more than 24"(61cm). Virtually all ceilings, walls, insulation, carpet, cushion and subfloor in the area are saturated.

Class 4 Water Loss: Specialty drying situations, focusing on bound water, involving wet materials with very low permeance/porosity (e.g., hardwoods, subfloor, plaster, brick, concrete, stone, crawlspaces). Typically, there are deep pockets of saturation that normally require very low specific humidity to dry.

Classes of Water: See *Class of Water Loss.*

Classification: See *Class of Water Loss.*

Clean Air Delivery Rate: See *CADR.*

Clean Water See *Category 1.*

Closed Drying System: System of drying where the structure is kept closed to outside air. Used because of security reasons and/or unfavorable outdoor air, or when access to outdoor air is prevented.

CMH: Cubic Meters per Hour - the metric measurement used for volume of airflow. Most often a measure of effectiveness in an air mover or axial fan. Can be used to determine the anount of ventilation in a structure. See also *CFM.*

Code of Federal Regulations: United States federally mandated practices and procedures administered by the Occupational Safety and Health Administration. Of particular interest to restoration contractors are: 29 CFR 1910, General Industry Standards and 29 CFR 1926, Construction Industry Standards.

Combination Drying System: System of drying where some outdoor air is used with dehumidification to accelerate drying. System can include flushing the air, negative pressurization and/or utilizing a small opening to the outside.

Commercial Refrigerant Dehumidifier: See *Standard Refrigerant Dehumidifier.*

Comply: An engineered wood material similar to plywood. Made from sheets or veneers of wood laminated to a particle board core. Is sometimes used as a subflooring material and can be found as a structural component in many furniture items.

Compression Extractor: Aggressive extraction tool that uses a combination of weight and vacuum seal to remove water from carpet and pad. Can be "ride on," self-propelled tools or manual, "drag" tools.

Compressor: The mechanical pump that moves refrigerant through a refrigeration system. Also referred to as a heat pump.

Concrete: A hard, strong construction material consisting of sand, gravel, broken stone, or slag in a mortar or cement (calcined limestone and clay).

Concrete Board: A panel made from concrete and fiberglass, commonly used a tile backing material.

Condenser Coil: A component in the basic refrigeration system where refrigerant gas is compressed, generating high temperatures. Heat leaves the refrigerant, causing it to condense.

Condition 1: From the IICRC S-520 mold standard, the condition of a structure and contents that indicates there is no active mold growth and only a normal presence of mold spores. (Structures are normally at Condition 1.)

Condition 2: From the IICRC S-520 mold standard, the condition of a structure and contents that indicates an increase in mold spores over normal levels, without active mold growth.

Condition 3: From the IICRC S-520 mold standard, the condition of a structure and contents that indicates an increase in mold spores over normal levels, in combination with active mold growth.

Confined Space: Area large enough for a person to enter, but not designed for continuous occupancy. See *Permit Required Confined Space.*

Conidia: Plural for conidium. An asexually produced fungal spore, produced on a conidiophore.

Conidiophores: A specialized fungal hyphae that produces conidia. See also *Conidia.*

Contamination: Refers to the presence of undesired substances, such as dirt, soils, microorganisms, chemicals that must be removed from the environment. See also *Category (1, 2, 3).*

Contents: A term used to describe items in a building that are not a permanent part of the structure, such as furniture.

Conventional Refrigerant Dehumidifier: See *Standard Refrigerant Dehumidifier.*

Crowning: Condition seen in hardwood flooring and other materials, caused by higher moisture content in the top of the material and lower moisture content in the bottom of a material. Often seen in hardwood flooring that has been dried beyond equilibrium moisture content.

CSA: Canadian Standards Association – the Canadian safety product label. Similar to UL.

Cubic Feet per Minute: See *CFM.*

CUL: Canadian Underwriters Laboratory. See *UL.*

Cupping: Condition seen in hardwood flooring and other materials, caused by high moisture content on the bottom side of the material and low or normal moisture content on the top side of the material.

Customer Briefing/Work Authorization Form: A document that summarizes key restoration activities that may be performed on any given restoration project; including information related to the unpredictable amount of time necessary to fully restore water damaged structures. The document is used as a communication tool to help building owners and occupants understand the restoration process. Also see the Appendix for a sample of this form.

Cut Pile: Type of carpet with fiber tufts that have been cut, leaving the ends of the fibers exposed. Most carpet is cut pile, and most cut pile is made of nylon fibers.

D

Damage: A physical change to a material that permanently alters its function or appearance. See also *Secondary Damage, Primary Damage, Pre-existing Condition.*

Data Logger: See *Data Recording Device.*

Data Recording Device: A device that captures data from an instrument over an extended period of time, creating a history of information. Data recorders are available to restoration contractors to capture information such as temperature, humidity, air pressure differentials and air particulate counts.

Deep Extraction: Refers to the extraction of the carpet underlay (cushion, pad). The deep extraction is performed with specialized extraction tools only. See *Vacuum-Sealed Extraction Tool, Weighted Deep Extraction Tool.*

Defrost: Mode of operation in refrigerant dehumidifiers. The dehumidifier goes into defrost mode when the cold coils freeze up and airflow through the unit is restricted. Defrosting is achieved by passive means (less efficient) or by hot-gas bypass (more efficient). Defrosting is often necessary in later stages of a water restoration, when the air is dryer.

Defrost Mechanism: A device that controls the defrost cycles in a refrigerant dehumidifier. See also *Defrost.*

Dehumidification: When combined with evaporation, dehumidification is one of the two ways to remove water from water damaged structures (besides extraction). Occurs when water molecules are *removed* from the air by condensation in a refrigerant dehumidifier, desiccation in a desiccant dehumidifier, or are sent *outside* the structure by air exchange.

Dehumidifier: A device used to remove water vapor from the air. Two types of dehumidification equipment are common in the restoration industry: Desiccant and Refrigerant. See also *Desiccant Dehumidifier, Refrigerant Dehumidifier, Heat and Air Exchange.*

Delamination: The separation of primary and secondary backings on a carpet. See also *Backing.*

Desiccant Dehumidifier: Dehumidifier that uses a desiccant medium to pull moisture out of the air. Generally requires more amperage than a refrigerant dehumidifier. Desiccant dehumidifiers can continue removing moisture down to 0 GPP (g/kg).

Dew Point: Temperature at which air can hold no more moisture. (Relative humidity at dew point is 100%.) The further a surface temperature is below dew point, the more condensation will occur.

Digital Psychrometric Calculator: Digital calculator that calculates GPP (g/kg), thirst and dew point when temperature and relative humidity are known. Also calculates LGR dehumidification and air mover requirements, given dimensions of the affected area and class of loss.

Direct Glue Carpet: A term for a carpet installation method, wherein the carpet is directly adhered to a substrate with no separate cushion system.

Disinfect: Destroy most organisms, leaving a surface free of infection. See *Sterilize* and *Sanitize.*

Disruptive Drying: Term used to describe restorative drying approaches that involve the removal or physical alteration of building materials. Requires the replacement or repair of building materials.

Dot System: Method of tracking moisture in walls and on hard surface flooring. Red, yellow and green dots are used to show the level of the moisture. Blue dots indicate materials scheduled for removal As moisture moves down a wall on successive days of drying, progress is indicated.

Drag Wand (weighted): Heavy carpet wand designed to remove water from carpet and pad.

Droplet Fogger: A device that aerosolizes small liquid particles, used to apply deodorizers.

Dry Bulb Temperature: Temperature of the air in Fahrenheit or Celsius. Used on psychrometric charts to determine the values of GPP (g/kg), vapor pressure and other qualities.

Drying Chamber: An area that in which temperature, humidity and airflow are manipulated to facilitate drying of wet materials.

Drying Pie: A term describing the three critical components of evaporation; Humidity, Airflow and Temperature; otherwise referred to as H.A.T.

Drying System: A term describing any of a number of methods for facilitating evaporation and humidity control in a restorative drying environment.

Dry Rot: A term for a group of fungi that invade and deteriorate the cellulose in wood, most commonly *Merulis lacrymans, Poria incrassata,* and/or *Serpula lacrymans.* The term *dry rot* refers to the fact that the fungi can grow when wood has a relatively low moisture content, between 20 and 30%. See also *wet rot.*

Dry Standard: Drying goal established in water restoration, determined by measuring the moisture content of unaffected materials in the environment – or of similar materials in similar buildings at the same time of year.

Dwell Time: Length of time a substance remains on a material. In biocides, dwell time indicates how long the substance needs to be on the surface to ensure effectiveness. Always follow biocide labels, industry standards and local laws when using biocides. Dwell time also relates to how long the water from the water source has been in contact with the materials in the structure. The longer the water has been in contact with the structure, the more difficult it will be to return the structure to pre-loss condition.

E

EMC: See *Equilibrium Moisture Content.*
Emergency Action Plan: A plan for fire reporting, emergency evacuation, operation of critical equipment during evacuation, accounting for staff, rescue operations and contact information for those that can provide additional information about the plan. Companies with greater than 10 employees are required to maintain an Emergency Action Plan in writing, according to OSHA 29 CFR 1910.38(b).
Endotoxin: A toxin that is confined inside a microorganism and is only released when the organism is broken down or dies. Endotoxins are primarily associated with particular species of bacteria. See also *Exotoxin*.
Engineered Wood: Wood products that have been created by laminating or adhering wood fibers, chips and/or veneers (i.e., plywood, OSB, particle board, MDF, LVL).
Enthalpy: Amount of energy in a system. In water restoration, enthalpy is typically the energy (in BTUs) contained in an air mass.
Environmental Protection Agency (EPA): An independent agency of the United States federal government that is aimed at reducing pollution and protecting the environment.
EPA: See *Environmental Protection Agency.*
Equilibrium: A theoretical state in which elements in nature have become equal and are no longer changing, such as air pressure and temperature.
Equilibrium Moisture Content (EMC): Condition where a material remains at the same moisture content because it has stabilized with the air around it. Evaporation and moisture absorption continue, but they are at the same rate–resulting in no overall change in moisture content.
Equilibrium Relative Humidity (ERH): Relative humidity of the air in a structure or in a confined space, after the air has come to equilibrium with the structural materials present. To determine a material's moisture content, using ERH, contain the material and a thermo-hygrometer inside a vapor barrier and give the material time to reach equilibrium with the air inside the containment. A normal relative humidity (of 50% or lower) inside the containment indicates that the material has reached a normal equilibrium moisture content.
Evaporation: When combined with dehumidification, evaporation is one of the two ways to remove water from structures (besides extraction). Occurs when water molecules change from liquid state into a vapor state.
Evaporative Cooling: Loss of heat energy in a material due to the evaporation of water from it.
Exfiltration: Process of air (and any water vapor or contaminants in that air) rushing out of an environment through small cracks and pores in a structure. Ex-filtration is the result of positive air pressure in an environment.
Exotoxin: A toxin that is secreted by a microorganism into the surrounding medium. *Exotoxins* are primarily associated with particular species of bacteria. See also *Endotoxin*.
Extraction: One of the two ways to remove water from water damaged structures (besides evaporation/dehumidification). Can be performed with a variety of tools connected to a vacuum unit.

F

Face Weight: The amount of fiber on the traffic side of a carpet. Higher face weight carpet has more fibers and/or taller fibers.
Fahrenheit: A scale for measurement of sensible heat, commonly used in the United States. The metric equivalent is Celsius.
Fall Protection: An OSHA regulated safety requirement. When work related activities are performed on elevated surfaces or where falling is a risk, protections are required by OSHA.
Filter Efficiency Rating: A rating that denotes the effectiveness of a filter. See also *HEPA, CADR.*
Fire Prevention Plan: An OSHA regulated safety requirement for the prevention of fire in the workplace.
Firewall: Wall type often seen in commercial structures that has an extra layer of material designed to slow down the spread of fire – often found near kitchens and mechanical closets. Extra monitoring and more aggressive or disruptive restoration may be required to return firewall to a pre-loss condition.
Flushing: Process of ventilating a structure to remove moisture laden air and replace it with drier air. Typically the air inside the structure is replaced with air from outside.
Foam Furniture Block: A small (approx 1" square) block made of high density foam; designed to separate furniture legs from wet flooring materials, thereby preventing continued water migration and secondary damage.
Forced Air Furnace: A heating device that employs an air stream to deliver heat. See also *Indirect Fired Furnace.*
Forced Air Heaters: A heating system that delivers heat through a direct air stream into a space. Caution must be taken when a forced air system is employed to ensure proper pressurization, circulation, and to prevent the buildup of carbon monoxide.
Free Water: Water contained in wood that is in the open spaces between wood cells and within cell cavities. Relatively easy to remove, compared to bound water.
Fungi: Plural for fungus. Fungi are the kingdom of living organisms that includes mold, yeast, rust, mildew and mushrooms. The *fungi* commonly encountered in water damaged structures are mold, a surface *fungus*.

G

Gas Monitor: Device designed to measure the amount of a specific or multiple gases in the air, such as carbon monoxide. Used whenever an unsafe gas might be present in the work area.
Gauge: Resistance across an electrical wire. Lower gauge indicates lower resistance (typically thicker wire). Extension cords must be of a lower gauge than the cord on the equipment to which they are carrying electricity.
Germination: To sprout or grow, to come into existence. Germination refers to the point at which spores begin to transform from a reproductive body to a living organism.
GFCI: Ground Fault Circuit Interrupter – when used properly with drying equipment, helps eliminate possible electrical problems. Can prevent electrocution by detecting a ground fault, and breaking the current.
G/KG: Grams per kilogram. 100 GPP ÷ 6.999 = 14.29 g/kg. See *Humidity Ratio.*
Glutaraldehyde: A water soluble liquid, $OHC(CH_2)_3CHO$, with antimicrobial properties. Glutaraldehyde is used in some deodorizing agents. See also *Aldehyde.*
GPP: Grains per Pound. See *Humidity Ratio.*
Grain Depression: Difference in GPP of the air entering and air leaving a dehumidifier – an indicator of dehumidifier performance.
Gray Water *Category 2.*
Ground Fault Circuit Interrupter: See *GFCI.*
Gypcrete: A cement material made of about 95% gypsum (a hydrated calcium sulfate mineral). Used as a flooring underlayment to add fire resistance and to insulate against sound.
Gypsum: A colorless or yellow mineral, $CaSO_4{\cdot}2H_2O$, commonly used in plaster of Paris, fertilizer and building products such as plaster and wallboard. See also *Wallboard.*
Gypsum Wallboard: See *Wallboard.*

H

Hardwood: Woods, such as oak or maple, that consist of much more cellulose material than softwoods. Hardwoods hold less free water than softwoods.

Hazard Communication Plans: See *HAZCOM*.

HAZCOM: An abbreviation for Hazardous Communication. Refers to the Hazardous Communication Standard published by OSHA.

Heat and Air Exchange: A method of humidity control wherein air is continuously exchanged between the affected interior of a structure and the outdoor environment. Heat is applied to the incoming outdoor air to reduce its relative humidity and promote evaporation. See also *Open Drying System*.

Heat Disorders and Health Effects: An OSHA document discussing the disorders related to exposures to high temperatures.

Heat Pipe: A patented energy exchanging system used in some low grain refrigerant dehumidifiers.

Heat Transfer Resistance: See *R-Value*.

HEPA: High Efficiency Particulate Air filter, which can remove 99.97% of all particles down to 0.3 microns. HEPA filters are used in a number of devices, including air scrubbers, negative air machines, air filtration devices and HEPA vacuums. See also *CADR*.

High Efficiency Particulate Air Filter: See *HEPA*.

Highly Recommended: Term used in the IICRC S-500 that indicates the generally accepted method or procedure.

Hot-Gas Bypass: Method of defrosting a refrigerant dehumidifier in which hot refrigerant from the hot coil (condenser) is allowed to flow directly into the cold coil (evaporator), thereby quickly defrosting the cold coil. Hot-gas bypass allows refrigerant dehumidifiers to pull moisture out of the air down to 33°F (.5°C) See *Defrost*.

HP: High Power. A designation commonly used for carpet dryers that contain more powerful motors (3/4 horsepower or 1 horsepower) compared to standard carpet dryers (1/2 horsepower).

Humidistat: A device that is used to regulate the operation of a dehumidification system. It monitors humidity in an area and regulates the power supply to a dehumidification device, thereby controlling the humidity in the target area.

Humidity: Moisture vapor in the air – one of the three elements of psychrometry that can be controlled, along with temperature and airflow, to influence evaporation rates. See also *Relative Humidity* and *Humidity Ratio*.

Humidity Ratio: Humidity in air, measured in Grains of water Per Pound of air (GPP), or in grams per kilogram (g/kg). Commonly confused with *Specific Humidity*, a measurement not used in the restorative drying industry.

HVAC: Heating, Ventilation and Air Conditioning. System in buildings that controls indoor environmental conditions, usually with a combination of heating, cooling and dehumidification.

Hydrogen Peroxide: A colorless, strongly oxidizing liquid (H_2O_2) with antimicrobial properties.

Hydrostatic Pressure: The force exerted by a fluid (water) on the surrounding environment when the fluid is at rest. It is the term for the pressure exerted by ground water on a building's substructure, such as a basement.

Hygrometer: Meter that takes readings of the relative humidity in the air.

Hygroscopic: Materials that readily absorb or release moisture based on the surrounding humidity.

Hyphae: Long, branching filaments that make up the greatest part of the fungi structure. See also *Mycelia*.

Hypochlorite: A class of compounds comprised of organic and inorganic salt, such as calcium hypochlorite, used as a bleaching agent.

I

IICRC: The Institute of Inspection, Cleaning and Restoration Certification.

IICRC S-500: Standard of care document for the water restoration industry, with several revisions including: 1st edition (1994), 2nd edition (1999), 3rd edition (2006). Available from the IICRC.

IICRC S-520: Standard of care document for the mold remediation industry. Published by the IICRC (2003).

Indirect Fired Furnace: A heating device that heats an air stream indirectly, ensuring that heated air is clean. Indirect fired furnaces add no byproducts of combustion to the heated air stream.

Industrial Hygienist: A person who has knowledge, practice and experience related to industrial hygiene and practices in this field. Industrial hygienists may be called upon to assist in the evaluation of the indoor environmental condition and to prepare remediation protocol in certain circumstances.

Infiltration: Process of air (and any vapor or contaminants in that air) rushing into an environment through small cracks and pores in a structure. Infiltration is the result of negative air pressure in an environment.

Infrared Camera: A device that gathers infrared light emitted by multiple surfaces. The amount of infrared light gathered is converted to a photo image, displayed by the device in colors that are visible to the human eye (not in infrared). Because infrared light emission is greatly influenced by temperature, the image displayed by the device is used to evaluate temperature differences across many materials simultaneously. Infrared cameras are also referred to as thermographic cameras, thermal cameras and thermal imaging devices.

Infrared Thermometer: A device that measures the amount of infrared light emitted by a surface. The amount of infrared light gathered by the device is converted to an estimated temperature of the surface measured.

Initial Extraction: Refers to the first step in extraction of wet carpet and underlay (cushion, pad). The initial extraction is most commonly performed with a light wand. See also *Light Wand, Deep Extraction*.

Initial Inspection Report: A summary of pertinent information gathered during the initial inspection; includes the time and date of initial inspection, water source(s), length of time structure was wet prior to inspection, types of material affected and other important loss information. Also see the Appendix for a sample of this form.

In Place Drying: See *Aggressive Drying*.

Inspection mirror: A device that allows for visual inspection of limited access cavities through the use of an extendable, adjustable mirror.

Inter-Air Dryer: Drying tool used for drying wall cavities and hardwood floors under positive or negative pressure. Has high air pressure but low volume of airflow.

Inter-air Drying Systems: See Inter-air Dryer.

Invasive Moisture Meter: Meter used to assess moisture within a structural material. Uses electrical contact between two metal pins to calculate moisture content. Calibrated on wood, and wood readings are reported as percent moisture content. Readings on other materials are reported as points of moisture or as a wood moisture equivalency.

J

K

Knee Kicker: Carpet installation tool used to disengage carpet in conjunction with an awl.

L

Laminated Veneer Lumber (LVL): An engineered wood product. LVL is created by layering wood veneers, adhered with waterproof adhesives. The final product is cut to form dimensional wood products for framing and load bearing structural supports. Also referred to as Structural Composite Lumber (SCL).

Laminate Flooring: A term used to describe engineered wood products used as a finished flooring material made to resemble hardwood.

Laser Particle Counter: Device that measures the number of particles present in an air mass. Some models are capable of measuring particle size and measuring numbers of particles at differing size ranges.

Latent Heat: The amount of heat absorbed or released by a substance (i.e., water) as it changes state. Also referred to as the heat of transformation.

Lead: A dense, pliable toxic metal that can be found in building materials as an antimicrobial (i.e., paint) or as a structural element where corrosives may be present (i.e., pipe, cable shielding, solder). Restorers are required by United States federal law (29 CFR 1926.62 and 29 CFR 1910.1025) to perform special procedures, including testing, before handling building materials that might contain lead.

Level Loop: Type of carpet in which the tufts are not cut (as in cut pile), and the loops of the tufts are all the same height.

LGR: See *Low Grain Refrigerant Dehumidifier.*

Light Wand: Common carpet wand, most often used for carpet cleaning. Can be used for some aspects of water extraction in water restoration. For deep extraction, a more aggressive tool must be used.

Lock Out/Tag Out: A process in which mechanical or electrical devices are disconnected and prevented from releasing stored energy, and in which the device is locked to prevent accidental startup.

Loss Assessment Report: See *Initial Inspection Report.*

Low Grain Refrigerant Dehumidifier (LGR): Uses the same basic technology as a standard refrigerant dehumidifier. An LGR uses a compressor and refrigerant to condense moisture on a cold coil. Air going through the dehumidifier is then warmed on the warm coils. The air is pre-cooled before the cold coils and pre-warmed before the warm coils by an air exchanger, heat pipe or heat siphon, resulting in higher efficiency. LGRs continue removing moisture in environments below 40 GPP (5.7g/gk).

Low Pressure Axial Fans: An air mover similar in appearance to an axial fan, but designed to deliver higher volumes of air at low power consumption rates. See also *Axial Fan.*

M

Makeup Air: Air that enters a structure by infiltration due to a negative pressure inside the structure. Makeup air needs to be controlled for moisture content, temperature and contamination.

Manometer: Device that measures air pressure differentials between two spaces. Manometers are available in varying pressure ranges and in varying pressure scales (i.e., inches of mercury, Pascals).

Manufactured Wood: See *Engineered Wood.*

Materially Interested Party: Individual or entity that has financial interest in the work being completed or in the items being restored. This may include but is not limited to: building occupants, building owner(s), insurance representatives, and mortgage companies.

Material Safety Data Sheet: See *MSDS.*

Material Vapor Pressure: A term used to describe the vapor pressure associated with moisture in a wet material. See also *Vapor Pressure.*

MC: (or %MC) See *Moisture Content.*

Mechanical Dehumidification: See *Dehumidifier.*

Medium Density Fiberboard (MDF): An engineered wood product where wood cellulose has been pressed and adhered to form a thick, dense product. Commonly found as baseboard, trim and as a structural component in some contents (i.e., entertainment centers, book cases).

Meter calibration: See *Calibration.*

Microbial Contamination: Contamination that includes microorganisms that could cause sickness, present in especially Category 2 and Category 3 losses.

Microbiological Volatile Organic Compound (MVOC): Volatile organic compounds produced by microorganisms as a byproduct of digestion. MVOCs are the odor associated with especially bacteria and fungus (mold).

Micron: A unit of measure. One micron is one millionth of a meter (there are one million microns in one meter).

Microorganism: Any organism that is of microscopic size, or is too small to see with the unassisted eye (i.e., bacteria, viruses, fungi).

Mitigating: The process of minimizing or reducing. Commonly used to describe initial restoration procedures that reduce the damage caused by abnormal water intrusion, such as water extraction, humidity control and furniture/contents protection.

Mitigation: Process of lessening the damage that would be caused in water damage. Prompt mitigation is the key to keeping the cost of restoration as low as possible. Mitigation prevents secondary damage and reduces dwell time, saving cost and time.

Moisture Content: Percentage of moisture in wood when compared to an oven-dried sample. Expressed in percentage of the dry weight.

Moisture Map: Drawing of the room or area that shows readings of moisture content, level in the structural materials. A shrinking area of moisture over successive days of drying indicates progress is being made on that structure.

Moisture Meter: See *Penetrating Moisture Meter and Non-Destructive Moisture Meter.*

Moisture Sensor: Device used to determine the horizontal migration of water in a carpet and pad – often a long stick with pin sensors at the end for ease of use, but sometimes a hand-sized unit for portability. Not to be used to check moisture content of baseboards, walls or non-porous flooring. Not to be used for determining if a structure has reached equilibrium moisture content.

Mold: Growth of microorganisms sometimes seen in water losses which sat for a long period of time or which were improperly dried. Prompt drying and cleaning is the best defense against mold.

MSDS: Material Safety Data Sheet. A document that contains important safety information for a substance such as its flammability, human exposure limits and chemical composition.

Multi-Level Loop: Type of carpet with tufts that have not been cut and which are different lengths, resulting in loops of different heights.

Must: Term used in the IICRC S-500 that indicates a mandatory practice procedure.

MVOC: See *Microbiological Volatile Organic Compound.*

Mycelia: Plural for mycelium. A mycelium is a mass of threadlike hyphae that make up the vegetative portion of fungi.

Mycotoxin: A toxin produced by fungus; such as aflatoxin, produced by the fungus Aspergillus.

N

National Electric Code: See *NEC.*

NEC: The National Electric Code. Published by the National Fire Protection Association, a nonprofit organization that develops consensus standards for building, fire and electrical safety.

Negative Air Machine: An air moving device used to evacuate air from a space, thereby creating a negative air pressure.

Negative Pressure: Condition created when a device is removing air from the inside of a structure directly to the outside. As more air is removed, negative pressure increases. Negative pressure is one cause of infiltration.

Noise Exposure: A potential safety hazard, regulated by OSHA. Information is available on permissible noise levels and lengths of time workers can be exposed to loud noise. Personal protection equipment is also available to limit the effects of noise exposure.

Non-Destructive Moisture Meter: Meter that uses radio technology or other means to determine the moisture content of a structural material without damaging it. Most are calibrated on wood, and wood readings are reported as percent moisture content. Readings on other materials are reported simply as Points. Some non-destructive meters have different settings for different materials.

Non-Invasive Moisture Meter: See *Non-Destructive Moisture Meter.*

Non-Penetrating Moisture Meter: See *Non-Destructive Moisture Meter.*

Nylon: A high-strength, flexible, resilient synthetic polymer fiber used in the construction of textiles such as carpet.

O

Olefin: A synthetic polymer fiber composed by weight of at least 85% ethylene, propylene, or other olefin units. Also referred to as polypropylene and common in the construction of textiles such as carpet.
Open Drying System: System of drying where outdoor air is used. Among other considerations, outdoor air should have a significantly lower GPP than the indoor air. See also *Heat and Air Exchange*.
Organic Compound: Any compound that contains carbon.
Oriented Strand Board (OSB): An engineered wood product manufactured from waterproof heat-cured adhesives and rectangular shaped wood strands arranged in cross-oriented layers, similar to plywood.
OSB (Oriented Strand Board): Building material often used in subflooring and sheathing applications. Made of large chips of wood held together with resin.
OSHA: Occupational Safety and Health Administration, a division of the United States Department of Labor.
Ozone: An unstable, poisonous, highly oxidizing allotrope of oxygen (O_3). It is used as to deodorize air and purify water. Ozone has antimicrobial properties.

P

Padding: Setting furniture and other contents items on blocks of foam, plastic or wood to help prevent secondary damage and provide a space for airflow underneath the item. Padding is also an alternate term for carpet cushion or underlay.
Parasite: An organism, especially fungus or bacterium, that derives its nourishment from other living organisms. Opposite of saprophyte.
Parquet Floor: A patterned wood inlay used to cover floors. Wood is made into tiles where the grain from each section of tile is perpendicular to the next.
Particle Board: An engineered wood product where wood fiber (sawdust sized) has been pressed and adhered to form a thick, dense product. Commonly found as an underlayment and as a structural component in some contents (i.e., entertainment centers, bookcases, shelving).
Pathogen: Any disease producing agent; especially viruses, bacteria or other microorganism.
Penetrating Moisture Meter: Meter used to assess moisture within a structural material. Uses electrical contact between two metal pins to determine the moisture content. Calibrated on wood, and wood readings are reported as percent moisture content. Readings on other materials are reported simply as Points.
Permeance: A measure of the ability for vapor to travel through a material. Permeance is measured in grains of water vapor, per hour, per square foot of material with a difference of 1.0 inches Hg of vapor pressure on opposing sides. See also *Vapor Barrier*.
Permit Required Confined Space: Confined space which contains a hazard of some kind and which causes a heath concern for employees. See *Confined Space*.
Personal Protective Equipment: See *PPE*.
Phenolic Compound (Phenol): A caustic, poisonous white-crystalline compound (C_6H_5OH) with antimicrobial properties. Also called carbolic acid.
Plain Sawn: The most common way in which wood is milled or cut. It results in boards that display the annual rings as flares on the flat, wide surface of the lumber; annual rings are nearly parallel to the cut. Plain sawing lumber produces more wood from each log processed, and therefore produces a lower cost product. See also *Quarter Sawn*.
Plaster: A mixture of lime or gypsum, sand and water, sometimes with fiber added, that hardens to a smooth solid and is used for coating walls and ceilings.
Plenum: A part of a heating or cooling system assembly to which the distribution ducting is joined.
Plywood: Building material used in all types of construction including subflooring, exterior sheathing and roof decking. Made of sheets of wood held together by resin.
Polypropylene: See *Olefin*.
Porous: A term that refers to a material or medium that contains pores, or holes. Highly porous substances are typically easily penetrated by vapor and aerosolized particulate.
Portable Generator: A device used to generate temporary power, typically through the use of a gasoline or diesel engine.
Positive Pressure: Condition created when a device is moving air from the outside of a structure directly to the inside of a structure. As more air is moved, the positive pressure increases. Positive pressure is a cause of ex-filtration.
Power Box: A temporary power distribution device most commonly used in the US and Canada to convert 230 volt power (i.e., stove outlets, dryer outlets) to 115 volt power.
PPE: Personal Protective Equipment. Refers to equipment such as boots, gloves, eye protection, body coverings, hard hats and respiratory protection equipment.
Pre-Existing Condition: Condition such as damage, delamination or pile crushing that was already present in the structure or contents before the water intrusion. All pre-existing conditions should be documented and called to the attention of all parties involved.
Pre-existing Contamination: A contaminant, such as microbial growth, in a structure that is not directly associated with the most recent water intrusion event. See also *Pre-existing Condition*.
Pre-Existing Damage: See *Pre-Existing Condition*.
Pressure Differentials: Refers to a difference in air pressure between two spaces. Where pressure differentials exist, air will move to balance the difference. See also *Infiltration, Exfiltration*.
Primary Damage: Damage to structure and contents that occurs because of contact with liquid water or contamination during the water damage.
Process Air: The air that has been dried by a dehumidifier. Used most commonly to refer to the warm, dry air produced by a desiccant dehumidifier.
Protozoa: Plural for protozoan. Any of a group of single celled organisms that are usually microscopic. *Protozoa* include amoebas, ciliates, flagellates, and sporozoans.
Psychrometric: Of or relating to psychrometry and its principles. See also *Psychrometry, Psychrometric Chart, Psychrometric Calculator*.
Psychrometric Calculator: Slide or dial type of psychrometric chart which calculates GPP (g/kg) if relative humidity and temperature are known. See also *Digital Psychrometric Calculator*.
Psychrometric Chart: Chart that shows the relationship between temperature, relative humidity, dew point, GPP (g/kg) and vapor pressure. If either temperature or relative humidity is known along with one other value, all other values can be calculated.
Psychrometry: The science and practice associated with "humidity," its evaluation, control and effect on materials.

Q

Quarter Sawn: A process used to saw lumber that results in annual rings perpendicular to the wide face of the board. This process produces lumber with better characteristics than other methods, but yields less usable material. Quarter sawn lumber is generally more expensive than wood sawn using other methods. See also *Plain Sawn*.
Quaternary Ammonium Compound (Quat): A compound derived from ammonium where hydrogen atoms have been replaced with organic groups; has antimicrobial properties. Quats are a common active ingredient in many antimicrobial products, both household and commercial.

R

RDC: See *Record of Drying Conditions*.

Reactivation Air: The air in a desiccant dehumidifier that has been used to remove collected water from the desiccant rotor. Reactivation air is typically hot and extremely humid, and is generally vented outside.

Recommended: Term used in the IICRC S-500 that indicates an advised or suggested method or procedure.

Record of Drying Conditions: Record of all temperature, relative humidity and GPP (g/kg) readings taken at a loss site. Includes readings for affected area, unaffected area, outdoors, dehumidifier output and any other pertinent readings (e.g., attic, crawl space, HVAC outlet, fresh air make up, etc.)

Refrigerant Dehumidifier: Can be either a standard refrigerant or low grain refrigerant. See *Standard Refrigerant Dehumidifier or Low Grain Refrigerant Dehumidifier.*

Relative Humidity: The amount of water vapor in air expressed as a percentage of the air sample's total holding capacity.

Remediation: The process of decontamination, cleaning and/or restoration when contaminants are present in significant quantity.

Respiratory Protection Plan: A written plan for respirator training and maintenance required by OSHA; mandated by 29 CFR 1910.134.

Restorative Drying: The process of mitigating water damage through the removal of excess water in building materials and contents.

Restoration Project Manager: See *RPM.*

R Factor: See *R-value.*

RH: See *Relative Humidity.*

RPM: Restoration Project Manager. A template, created in Microsoft® Excel®, for capturing all information regarding the restoration project. See CD-ROM for this template.

R-Value: The resistance to heat energy transfer that an insulating material has. The higher the R-value, the more resistant the material is to heat transfer.

S

S500: See *IICRC S-500.*

S520: See *IICRC S-520.*

Safety plan: A documented plan generated by the restoration firm that outlines specific practices and procedures for identifying and eliminating safety hazards. Preparation of a documented safety plan is a necessary part of compliance with United States federal law; see also *Code of Federal Regulations.*

Sanitary: A term describing the level of cleanliness in a space or area, denoting that the level of filth and pathogens do not pose a substantial risk to human health.

Sanitize: Return a material or structure back to a normal healthy condition after a contamination of some kind. See *Sterilize* and *Disinfect.*

Sanitizers: A term used by the EPA to describe the lowest level of biocidal activity.

Saprophyte: An organism, especially fungus or bacterium, that derives its nourishment from dead decaying organic matter. Opposite of *Parasite.*

Saturation: Point at which materials or an air mass can no longer absorb more moisture. Wood fiber saturation is 30%; moisture contents above 30% indicate free and potentially surface moisture. Air will achieve saturation at 100% relative humidity, or dew point temperature.

SCBA: See *Self Contained Breathing Apparatus.*

Secondary Backings: See *Backing.*

Secondary Damage: Damage to structure and contents that occurs because of high humidity, loss of structural integrity, microbial contamination, etc. The purpose of water loss mitigation is to prevent secondary damage.

Self Contained Breathing Apparatus: A respiratory protection device that includes its own supply of breathable air.

Sill Plate: Dimensional lumber attached to the top of a building foundation. Rim joists rest on the sill plate.

Softwood: Woods such as pine or fir. Softwoods consist of much less cellulose material than hardwoods, and also hold more free water.

Sole Plate: See *Base Plate.*

Source: Water intrusion point of origin. Depending on the source, different procedures will be used to restore the structure to pre-loss condition.

Specific Humidity: a measurement of water vapor weight in a given volume of air. Often confused with GPP (g/kg). See *Humidity Ratio.*

Sporangium: A structure in which spores are produced. Also called a spore case.

Spore: A small, usually single celled reproductive body that is highly resistant to dry conditions, temperature and other adverse conditions. Spores are produced by fungi, bacteria and other organisms.

Standard Refrigerant Dehumidifier: A dehumidifier that employs the process of condensation on a cold surface. Refrigerant dehumidifiers use similar components to air conditioning units, where cold and hot surfaces are created with gases in a sealed coil system.

Sterilize: Destroy all living organisms on a material or in a structure. The highest level of biocidal activity. See *Sanitize* and *Disinfect.*

Stretched-In Carpet: A term for a carpet installation method, wherein the carpet is stretched in over pad. Carpet is held onto tackless strip around the perimeter. See also *Carpet.*

Subfloor: Floor just above the support beams or joists. The subfloor supports flooring material such as carpet and pad, tile or vinyl flooring.

Sublimation: Process of water going from a solid (ice) phase to a gas (vapor) phase without becoming a liquid in the transition.

Submersible Pump: A device used to remove standing water. Many submersible pumps are capable of removing standing water to levels as low as ¼" in depth.

Surface Water: Moisture that resides on the surface of a material. Generally, this moisture will leave the material readily and is visible. See also *Free Water, Bound Water.*

Swelling: Increase in size of a material due to the absorption of water.

System, Balanced: See *Balanced Drying System.*

T

Temperature: Measure of the heat quantity in the air. Temperature is a direct measure of the movement of molecules of a substance. The higher the temperature, the faster the molecules move. Temperature, along with humidity and airflow, is one of the three components of psychrometrics that can be controlled in a water restoration.

Temporary Power Distribution Panels: A device used to establish temporary power supplies in common voltage from other power sources, such as high voltage power drops from utility lines or utility outlets such as clothes dryers and ranges (stoves).

Tested Airflow: The actual amount of airflow from a device, such as an air mover, which represents real airflow more accurately than rated airflow.

Thermal Capacitance: The ability of a substance to hold heat.

Thermal Siphon: A device that exchanges energy in a refrigeration system to increase water removal performance and energy efficiency. Used in some low grain refrigerant dehumidifiers.

Thermodynamics: The science that deals with changes in energy. Principles of "hot to cold," "wet to dry" and "high pressure to low pressure" come from the science of thermodynamics.

Thermo-hygrometer: Meter that takes a reading of temperature and relative humidity. Some thermo-hygrometers also calculate grains of water per pound of air.

Thirst: Difference between the current GPP (g/kg) of air and the maximum GPP (g/kg) at that same temperature. Higher *thirst* promotes evaporation. Air at 75°F (24°C) and 30%RH has a thirst of 92 GPP (13.1g/kg).

Thirsty: See *Thirst.*

Time Temperature Defrost Systems: A defrost control system that uses ambient air temperatures to remove frozen water from refrigerant dehumidifier coils. See also *Hot Gas Bypass, Defrost.*

Ton of Air Conditioning: : Amount of energy needed to change one ton of 32°F(0°C) ice to liquid water, or 12,000 BTUs. One ton of air conditioning is equal to 12,000 BTUs per hour.

Toxin: A poisonous substance, usually a protein, that is capable of causing disease when introduced to the body or tissues but is also often capable of inducing antibodies and antitoxins. See also *mycotoxin, endotoxin, exotoxin.*

U

UL certified: Certification by Underwriters Laboratories that shows a piece of equipment or a component of the equipment has been certified safe by UL. Any modifications from original product design voids UL certification.

Ultra Low Volume (ULV): A term used to describe the application of a liquid agent to air through aerosolizing of small droplets, or a fine mist; Ultra Low Volume Fogger.

ULV: See *Ultra Low Volume.*

Underlay: A cushioning material installed under stretched-in carpet made from urethane, rubber, felt or other similar materials. Also referred to as a carpet pad or cushion.

V

Vacuum Compression Extractor: See *Vacuum-Sealed Extraction Tool.*

Vacuum-Sealed Deep Extraction Tool: A tool designed for extracting water from carpet and underlay (pad, cushion). Vacuum-sealed tools remove water from the carpet and underlay predominantly by means of vacuum, generating airflow through the underlay to carry water up and into the tool.

Vapor Barrier: material with a permeance of 1.0 or less. See also *Permeance.*

Vapor Diffusion: Movement of water vapor through a material without changing state to a liquid and having no effect on moisture content of that material.

Vapor Pressure: Pressure of water vapor on surrounding surfaces and objects, measured in inches of mercury. Directly related to GPP (g/kg).

Vapor Retarder: Any material with a permeance factor of less than 1.0 perm. See *Permeance.*

Virus: Any of a group of agents that consist of DNA or RNA surrounded by a protein layer. Viruses cannot replicate without a living host and are therefore not considered to be living. They are however considered "live" when they are capable of replicating and causing disease.

Volatile: Evaporating readily at normal temperatures and pressures, is easily vaporized.

Volatile Organic Compound: Any compound containing carbon that is easily vaporized or evaporated at room temperature and normal pressures.

Volt: See *Voltage.*

Voltage: Electrical potential or "push" measured in volts. Residential voltage is typically 110-120 volts (US), 230 volts (EU). Combining two residential lines of the same polarity results in doubling the voltage.

Vinyl Composition: Composed of or containing vinyl. Common in hard surface flooring materials such as tile or sheet goods (Vinyl Composition Tile, Vinyl Sheet Goods).

W

Wallboard: A product made of gypsum and paper commonly used on interior wall and ceiling surfaces. Also referred to as Sheetrock®, Drywall® and plaster board.

Wall Plate: See *Base Plate.*

Was-Wood: See *Engineered Wood.*

Water Activity: Moisture available at the surface of a material, expressed as a 100th of the ERH (Equilibrium Relative Humidity). The symbol for water activity is Aw.

Water Damage Customer Information Form: Information the restorer provides to let the customer know about the process that will be followed during the restoration.

Watt: Unit of power that can be calculated by multiplying voltage x amperage.

Weep Hole: Hole drilled or punched through a structural material to let water drain from a cavity. (Be extra cautious when placing weep holes overhead.)

Weighted Deep Extraction Tool: A tool designed for extracting water from carpet and underlay (pad, cushion). Weighted tools remove water from the carpet and underlay predominantly by means of weighted compression, forcing water to move out of underlay, through carpeting and up into the tool.

Wet Rot: A term for a group of fungi that invade and deteriorate the cellulose in wood, commonly *Conidiophora cerebella.* The term wet rot refers to the fact that the fungi prefers damp wood, with a moisture content greater than 30%. See also dry rot.

Wood Moisture Equivalency: Reading obtained on a moisture meter when checking a non-wood material. Measured in Points.

Work Authorization: Instructions and authorization for the restoration of the property. Do not start work until this form has been signed. See *Appendix for an example.*

Woven Carpet: Refers to a carpet that is made using a weaving process. The face and back of a woven carpet are made at the same time as yarns are woven together. It is more common to find organic materials in a woven carpet than in other types of carpet. See also *Carpet.*

X

Y

Z

Select Bibliography

PUBLICATIONS:

Gatley, Donald (Author). *Understanding Psychrometrics.* Atlanta, GA: American Society of Heating, Refrigerating and Air-Conditioning Engineers, Inc., 2002.

Parsons, Robert (Editor). *ASHRAE Handbook-Fundamentals I-P Edition.* Atlanta, GA: American Society of Heating, Refrigerating and Air-Conditioning Engineers, Inc., 1997.

Moffet, Patrick (Author). *Managing Microbial Problems in Buildings.* Huntington Beach, CA: Environmental Management & Engineering, Inc., 1st edition, 2003.

S520 IICRC (Publisher) *S520 Standard and Reference Guide for Professional Mold Remediation.* Vancouver, WA: Institute of Inspection, Cleaning and Restoration Certification (IICRC), 1st edition, 2003.

S500 IICRC (Publisher) *S500 Standard and Reference Guide for Professional Water Damage Restoration, 2006.* Vancouver, WA: Institute of Inspection, Cleaning and Restoration Certification (IICRC), 1st edition (1994), 2nd edition (1999), 3rd edition (2006).

WEBSITES:

Occupational Safety and Health Administration	www.osha.gov
Occupational Safety and Health Administration	www.osha.gov/dts/shib/shib101003.html
The Wood Flooring Manufacturers Association	www.nofma.org
ASTM International	www.astm.org
Dri-Eaz Products, Inc.	www.dri-eaz.com
United States Environmental Protection Agency	www.epa.gov

Additional resources available include the following documents, as recommended additional reading:

- *Suggested Guidelines for Remediation of Damage from Sewage Backflow into Buildings*
- *IICRC S500 Water Damage Restoration Standard and Reference Guide, Third Edition*
- *Recommended Immunizations for Sewage Workers, California Morbidity*

Index

Index

Index